Summary of Occasional Paper

CRISIS

GW01454158

Will there be an economic crisis in 1975?—or 1976, 1977 . . . ?
What form will it take—a run on sterling? hyper-inflation? mass
unemployment? shortages of food? Will it be a last straw or a
turning point?

What are the *immediate* causes? Monetary expansion or trade
union pressure? Soaring oil or raw material prices?

What are the *ultimate* causes? Over-full employment? High
government expenditure? Discouragement of enterprise? Trade
union monopoly?

Are government measures—overseas borrowing, subsidies to
uneconomic firms, maintenance of jobs—wise palliatives or short-
sighted alleviation of symptoms? Would lower wages combat
inflation?

What are the remedies? Should the unions be persuaded to
restrain their bargaining power by 'incomes policies', or be de-
prived of their monopoly power by law?—or should trade unionists
suffer financial penalties if they extract 'excessive' pay? Can unemployment stop inflation? Should 'fine-tuning' be replaced by
steady change in the money supply?

What should government do *now*? Reduce consumption?
Cut taxes on profits or reduce interest rates to stimulate investment?
Reduce borrowing from abroad? Relax or end price controls?
Raise charges for government welfare or other services? Prevent
hardship by a reverse income tax? Prune the bureaucracies,
national and local? Introduce indexing of tax allowances, incomes,
savings, etc.? Reform the wage-fixing machinery?

Can the British Parliamentary system resist abuse of the
economy by political parties for electoral advantage?

16 authors of varying schools of thought provide a unique feast
of intellectual stimulus for all concerned for the future of Britain.

Occasional Paper Special (43) is published (price £1.50) by

THE INSTITUTE OF ECONOMIC AFFAIRS

2 Lord North Street, Westminster, London SW1P 3LB

CRISIS '75..?

SIR JOHN HICKS	University of Oxford
HENRY PHELPS BROWN	University of London
JAMES E. MEADE	University of Cambridge
LORD KAHN	University of Cambridge
HENRY SMITH	Formerly Ruskin College, Oxford
SIR ALEC CAIRNCROSS	University of Oxford
PETER M. OPPENHEIMER	University of Oxford
WILFRED BECKERMAN	University of London
PAUL BAREAU	'University of Fleet Street'
MALCOLM R. FISHER	University of Cambridge
RALPH HARRIS	Formerly St. Andrews University
E. VICTOR MORGAN	University of Reading
ALAN A. WALTERS	University of London
MICHAEL PARKIN	University of Manchester
SAMUEL BRITTAN	University of Oxford
HARRY G. JOHNSON	University of Chicago

Published by
THE INSTITUTE OF ECONOMIC AFFAIRS
1975

First published January 1975

by

THE INSTITUTE OF ECONOMIC AFFAIRS

© The Institute of Economic Affairs 1975

All rights reserved

SBN 255 36066-5

Printed in Great Britain by
GIBBONS BARFORD, WOLVERHAMPTON
Set in Monotype Plantin

Contents

Preface
A CONCORDAT

IEA *Occasional Papers* are designed, *inter alia*, to present to students and teachers of economics, and to laymen interested in economic thinking, material originally delivered or published for specialist groups of listeners or readers. They are also used to present material that does not easily fall within the *Hobart Papers*, *Hobart Paperbacks* or other IEA series.

In November the Institute was approached by several readers about making a contribution to the developing discussion of the supposedly impending economic 'crisis'. Apprehension was spreading in the press and elsewhere about the risks not only to living standards but also to democratic institutions.

The Institute has tended to concentrate on long-term or medium-term and avoided short-term issues, where economists are not characteristically at their best, where economics as a science earns little repute, and where deficiencies of forecasting are often painfully plain: in the British economy frequent changes in government policy, prompted by over-reaction to underlying structural developments, are themselves de-stabilising and unpredictably arbitrary and make forecasting precarious (cf. Chart 7, p. viii). But in view of the growing public debate and anxiety, and the disproportionate access of newcomers and non-economists to economic thinking of the various schools, the Institute has invited economists of a wide range of approach and persuasion to write short essays on how they saw a crisis in 1975. They were originally invited to write 1,000 to 1,500 words by mid-December but, when several asked for more space and time, the length was extended to 2,000 words and the date to 1 January (for publication on 31 January). The Institute would like to thank them for their response; and it is especially indebted to Mr Michael Solly, its Publications Manager, and to the printer and his staff for a remarkably expeditious and pleasing job of printing.

The result – this *Occasional Paper Special* – is a rare, possibly a unique, assembly of essays by some of the cream of Britain's economic talents that includes long-famous names and fast-rising

[5]

younger scholars in seven (eight!) universities: 14 current or former University teachers and two of Britain's most distinguished economic commentators, one associated with an Oxford college, the other deserving, in a just world, to be dubbed Distinguished Professor of Applied Economics in the University of Fleet Street who, since his apprenticeship with Sir Oscar Hobson on the liberal *News Chronicle*, has taught millions the essentials of economics with a command of English and a verbal grace native economists must envy. Some of the essayists have differed strongly among themselves, but most here convey a sense of alarm and some sound a note of urgency that the British economy requires new thought and unpalatable action if it is not to collapse or disintegrate, if not in 1975 within a few years.

*   *   *

A superficial observer might repeat the familiar taunt that, since no two economists can agree with each other, economists cannot offer guidance to a troubled world. There are reasons why social scientists differ more than natural scientists, but the 16 contributors to this IEA *Paper* mostly congregate around two or three broad themes in the cause and cure of inflation. Five or six are 'monetarists', more or less. Most of the remainder are 'institutionalists' who find the ultimate cause in 'real', structural or institutional conditions: vested interests, notably monopoly trade unions, or intellectual influences, notably the commitment to full employment and high government expenditure. And several concentrate on one or two 'institutional' elements: Lord Kahn emphasises the impact of accelerating oil prices, Mr Henry Smith the rate of interest and industrial investment, and Mr Samuel Brittan the political economy of inflation.

The differences may be even less than these sub-divisions suggest. One contributor, Mr Ralph Harris, attempts a reconciliation of the monetarist and institutionalist (monopoly union) approaches. It may be that, to emphasise the element to which each approach attaches *primary* importance, both understate the element emphasised by the other. The monetarists may understate the institutional pressures on government, especially in Britain, to inflate: and the institutionalists may understate the money supply as the mechanism or *instrumentality* of inflation. Here may lie the economic concordat. If the monetarists would concede that governments are under severe industrial and ideological pressure to inflate the money supply, and the institutionalists acknowledge that inflation cannot proceed by 'cost-push' unless government inflates the money supply (which itself intensifies union power), the two approaches would move much closer together. A larger measure of agreement might then be reached on 'government-push' as the central explanation, cause and culprit. The debate would, at least, have been reduced to the meaning of 'cause' (the money supply as the proximate cause, institutions as the ultimate causes) and to judgement on the timing and effectiveness of 'incomes policies' as a way of reducing the transitional unemployment accompanying monetary discipline. When the differences of 'political' judgement that divide economists are seen to be less important than the basic economic insights that unite them, the taunt that no two economists, or schools of economists, ever agree about anything might then be seen to be insubstantial, and economics might regain some of its authority and respect.

If it seemed that economists of all or most schools were more or less agreed on *diagnosis*, the judgement on timing and on the relative roles for monetary control and 'incomes policy' would then be seen to be *political*, in which realm economists do not claim special insight or expertise. And the blame for failure would then fall where it should have fallen: on the politicians for misjudging the reactions of the people as employees, trade unionists, consumers or investors to changing prices, changing pay, and, not least, changing jobs. Even the difference of view between economists who believe that monopoly unions should or could be prevailed upon to abdicate or restrain their monopoly power and those who believe that monopoly power should be withdrawn by Parliament, insofar as it was conferred by law, would then be seen as a judgement on political probability rather than as a conflict in economic theory.

* * *

Within 2,000 words, more or less (Professor Harry Johnson had completed his contribution in the USA before he knew of the additional length), the contributors have not been able to explain their analyses and judgements fully: readers must refer to their published work for further documentation.[1] Laymen who feel unqualified to judge between alternative economic theses may have to judge them by events in 1975 and beyond. Whether or not the expectations

[1] The biographical notes, pp. 11-16, list their main writings.

prove true, laymen will find them intriguing and teachers and students of economics must find them a feast of stimulus in their academic studies as a wider range of views than those to which they may have easy access.

The arrangement of the essays presented a problem. The conventional but analytically arbitrary alphabetical order would have confronted readers with a succession of arguments that had no apparent logical sequence and that seemed to jump erratically from one interpretation to almost the opposite and back again. In conformity with the view that the fundamental analytical insights shared by economists of all schools carry more authority, and are more significant for the layman, than the differences in value-judgements which divide them, a rough shot has been made to arrange the essays according to their broad emphases, although (with the exception of the 'monetarists', who also differ in degrees) it is not intended to attach a label to each. Pride of place is given to the Nobel Prize-winner, Sir John Hicks, a course of which it is supposed all the other contributors would approve. The essays following seem to emphasise one or other or several of the 'real', institutional elements; the 'monetarists' are grouped some three-quarters of the way down the line; and the roll ends with a contribution that emphasises the constitutional environment if the economy is to survive and master inflation and with the only essay from an economist overseas.

*　　　*　　　*

The authors were supplied with an informal framework or dummy 'questionnaire'[1] but were not required to stay closely to it; some went

---

[1] Dummy questionnaire:
1. Other things being equal, is Britain likely in 1975 to confront a worsening crisis that will call for significant changes in Government policy?
2. What are the principal reasons for the present economic crisis:
   (a) monetary policy
   (b) trade union pressure
   (c) international developments, including oil
   (d) other failures of post-war governments – acts of commission and omission
   (e) other reasons
3. What form do you see the crisis taking in 1975?
4. What would be the likely consequences of leaving present government policies broadly unchanged?
5. How far will the developments you foresee make necessary a change in the tactics or the strategy of economic policy?
6. Are there any other changes which would be helpful in overcoming the danger of worsening crisis?

beyond it; what they wrote is reproduced here to show how they interpret recent developments and envisage required policies. No doubt some readers may wish to have seen some aspects developed further and others added. Not least, if government expenditure is to be cut, as is agreed on all (or most) sides, it would seem that, to have *early* effect, charges for hitherto free or almost 'free' government services, not least welfare, both national and local, could be introduced without hardship or injustice and with more effect on inflation than many other reforms discussed in the press. Politicians and officials in local as well as national government may have to be prepared for more radical policies than they are accustomed to contemplate. But it may be that the policy would find a large measure of public approval, as indicated in the series of studies conducted for the Institute and reported in *Choice in Welfare* 1970 and *Choice in Housing* 1968; if so, new reforms could be pursued with wide political agreement in reducing government expenditure, without which inflation may remain intractable.

The 'crisis' is not only an economic failure that can be analysed by economic theory but, as Mr Samuel Brittan emphasises, possibly even more a political crisis to be analysed by political economy. However much or little economists agree or differ on technical analysis, a truth that stands out above all else is that in practice 'public policy' can be a comforting, question-begging phrase that conceals the realities of everyday government decisions made by electorally-motivated politicians guided by fallible officials informed, or misinformed, by incomplete, misleading, and out-of-date statistics. Thinkers of all schools, Marxist, Keynesian and some liberal sub-schools, make the assumption that government can and will conscientiously apply the best economic advice. The classical economists knew better. In recent years their realism has been revived with new insights by Professors James Buchanan and Gordon Tullock of the Virginia School (developing earlier work by A. C. Pigou and Professor Paul Samuelson) that has evolved the new 'economic theory of politics' in which little remains of the conventional assumption that government is necessarily concerned with 'the public interest'.[1] If the crisis is seen as the culmination of errors in post-war economic policy, the main lessons may be, as emphasised

---

[1] Professor Gordon Tullock is working on a Hobart Paper, *The Economics of Democracy*, to be published in the autumn.

by Mr Harris, that government, inhibited by 'inbuilt imperfections', should not be asked to do what it cannot do; and that the best hope for the future lies in recognising the limitations of government and in developing new understanding and new attitudes by the public.

Some economists may also wish to emphasise the difference between the changing terms of trade in commodities and in factors of production. Others may wonder whether the discussion is about the dangers of inflation or of deflation. Some readers may wonder if a crisis will develop when the small saver wakes up and realises he earns a negative yield and is losing real capital value and withdraws his savings at an accelerating rate. If there are complaints of inefficient industrial management the fault may lie with government, which has failed to make the economy more competitive. Here industry is hardly likely to learn from the Civil Service.

*   *   *

An editorial note must be added. IEA *Papers* are the outcome of sometimes lengthy collaboration between IEA staff and authors. These 'crisis' essays were commissioned, written, assembled, typed and printed under pressure on all sides and without time for the normal editorial processes. Several authors were abroad, in Paris and Rome, or preparing to go abroad, others were preoccupied with other work required to be completed by the end of the year, and so on. Several essays were delivered within hours of arrival of the printer's emissary from Wolverhampton, and one arrived on 8 January after he had departed. Sources, references, allusions and documentation generally could not always be discussed with authors as is usual with IEA *Papers*. The side-headings and charts were added by the IEA staff.

The Institute's constitution requires it to dissociate its Trustees, Directors and Advisers from the analyses and conclusions of its authors. This formula is repeated here; several of the essayists are formally associated with the Institute but write in their personal capacity.

8 January, 1975   ARTHUR SELDON

The Authors

(in order of essay sequence)

SIR JOHN HICKS, Kt., FBA. Born 1904, educated at Clifton College and Balliol College, Oxford. Lectured at LSE 1926-35; Fellow of Gonville and Caius College, Cambridge, 1935-38; Professor of Political Economy, University of Manchester, 1938-46; Official Fellow, Nuffield College, Oxford, 1946-52; Drummond Professor of Political Economy, University of Oxford, 1952-65; Fellow of All Souls College since 1952; Nobel Memorial Prize for Economics (jointly), 1972. Knighted 1964; FBA 1942. Publications include: *The Theory of Wages* (1932, revised edn. 1963); *Value and Capital* (1939); *The Social Framework* (1942, 4th edn. 1971); *A Contribution to the Theory of the Trade Cycle* (1950); *Essays in World Economics* (1959); *Capital and Growth* (1965); *Critical Essays in Monetary Theory* (1967); *A Theory of Economic History* (1969); *Capital and Time* (1973); *The Crisis in Keynesian Economics* (1974).

HENRY PHELPS BROWN, FBA. Born 1906, educated at Taunton School and Wadham College, Oxford. Fellow of New College, Oxford, 1930-47; Royal Artillery, 1939-45; Professor of the Economics of Labour, University of London, at the London School of Economics, 1947-68, now Emeritus Professor. One of the 'Three Wise Men' in 1959-61, and a founder-member of the National Economic Development Council, 1962-66. Publications include *The Framework of the Pricing System* (1936); *A Course in Applied Economics* (1951); *The Growth of British Industrial Relations* (1959); *The Economics of Labour* (1963); *A Century of Pay* (1968).

JAMES EDWARD MEADE, CB, FBA. Born 1907, educated at Malvern College, Oriel College, Oxford, and Trinity College, Cambridge. Fellow and Lecturer in Economics, Hertford College, Oxford, 1930-37; economic section of the Cabinet Offices, 1940-47; Professor of Commerce, London School of Economics, 1947-57; Professor of Political Economy, University of Cambridge, 1957-68; President, Royal Economic Society, 1964-66; President, Section F, British Association for the Advancement of Science, 1957; Governor, NIESR, since 1947. Publications include: *Planning and the Price*

[11]

Mechanism (1948); *Efficiency, Equality, and the Ownership of Property* (1964); *Principles of Political Economy* (Vol. 1: 1965, Vol. 2: 1968, Vol. 3: 1972); *The Theory of Indicative Planning* (1970).

RICHARD FERDINAND KAHN, CBE, FBA. Born 1905, educated at St. Paul's School, London, and King's College, Cambridge. Professor of Economics in the University of Cambridge, 1951-72, and a Fellow of King's College. Temporary civil servant in Government Departments, 1939-46. Appointed CBE in 1946, FBA in 1960; created a Life Peer in 1965. Published works include *Selected Essays on Employment and Growth* (1973), and articles in the learned economic journals. A literary executor of J. M. Keynes.

HENRY SMITH started life as an errand boy at $13\frac{1}{2}$, went to Christ Church, Oxford, with an extra-mural scholarship at 25: First-Class Honours in PPE and a three-year post-graduate scholarship; Lecturer in Business Finance at Liverpool University; Resident Economics Tutor at Ruskin College, Oxford, of which he was Vice-Principal, 1947-70. Appointed to the Civil Service Arbitration Tribunal after the war, resigned 1964. Wage arbitration in Gibraltar and Hong Kong, and expert witness for the Registrar before the Restrictive Practices Court. Member of the Consumers' Advisory Committee of the Agricultural Marketing Boards. Publications include: *Retail Distribution: A Critical Survey* (1937); *The Economics of Socialism Reconsidered* (1962). For the IEA he has written *A Prospect of Political Economy* (1968), a Hobart Paper (No. 18), *The Wage Fixers* (1962, reprinted with a Postscript in *Freedom or Free-for-All?*, 1965), a Research Monograph (No. 6), *John Stuart Mill's Other Island* (1966), and an essay in IEA Readings No. 6, *Inflation and the Unions* (1972). IEA Advisory Council.

SIR ALEC CAIRNCROSS. Born 1911, educated at the Universities of Glasgow and Cambridge. Has alternated over the past 40 years between academic life and government employment. In 1961 he succeeded Lord Roberthall as Economic Adviser to the Government; in 1964 became the first Head of the Government Economic Service. Resigned in 1969 to become Master of St. Peter's College, Oxford. Has been President of the Royal Economic Society, the Scottish Economic Society and the British Association for the Advancement of Science. In 1972 he was elected Chancellor of the

University of Glasgow. Publications include *Introduction to Economics* (1944); *Factors in Economic Development* (1962); *Essays in Economic Management* (1971); *Control over Long-term International Capital Movements* (1973).

PETER MORRIS OPPENHEIMER. Born 1938, educated at Haberdashers' Aske's School, Hatcham, and The Queen's College, Oxford (1st class PPE); Student and Tutor in Economics, Christ Church, Oxford University, since 1967; Resident Fellow, Nuffield College, 1964-67. Publications: contributions to *Bulletin of the Oxford University Institute of Economics and Statistics*; *Journal of Money, Credit and Banking*; *Scottish Journal of Political Economy*; bank reviews.

WILFRED BECKERMAN. Born 1925, educated at Ealing County School and Trinity College, Cambridge. Lecturer in Economics, University of Nottingham, 1950-52; worked for the OEEC/OECD in Paris, 1952-61; National Institute of Economic and Social Research, 1962-63 (Member of the NIESR Executive Committee since 1973); Fellow of Balliol College, Oxford, 1964-69; Economic Adviser to the Board of Trade, 1967-69; Professor of Political Economy, University of London, and Head of the Department of Political Economy at University College, London, since 1969. Member of the Royal Commission on Environmental Pollution, 1970-73. Publications include: (with associates) *The British Economy in 1975* (1965); *International Comparisons of Real Incomes* (1966); *An Introduction to National Income Analysis* (1968); (ed. and contributor) *The Labour Government's Economic Record* (1972); *In Defence of Growth* (1974).

PAUL BAREAU. Born in Belgium; educated at Dulwich College and the London School of Economics. Most of his working life has been devoted to economic journalism. Editor of the *Statist*, 1961-67. Economic Adviser to the International Publishing Corporation and Economic Consultant to Barclays Bank. Secretary, Political Economy Club (founded by David Ricardo in 1821).

MALCOLM ROBERTSON FISHER. Born in New Zealand, 1923; educated at Auckland Grammar School and the Victoria University of Wellington; for four years Lecturer in Economics at University of Auckland; studied for a Ph.D. at Cambridge, 1950-52, and thereafter Research Officer at the Oxford University Institute of Statistics; since 1956 University Lecturer in Economics and Fellow and

[13]

Director of Studies in Economics, Downing College, Cambridge. A specialist in applied econometric studies, particularly consumption of agricultural commodities, personal saving and business financing, and more recently the economics of labour. Publications: *The Economic Analysis of Labour* (1971); *Measurement of Labour Disputes and their Economic Effects* (1974); contributions to learned economic journals. The IEA published his *Macro-Economic Models* (Eaton Paper 2, 1964).

RALPH HARRIS. Born 1924, educated at Tottenham Grammar School and Queens' College, Cambridge. Lecturer in Political Economy at St. Andrews University, 1949-56. General Director of the Institute of Economic Affairs since 1957. Author (with Arthur Seldon) of *Hire Purchase in a Free Society*, *Advertising in a Free Society*, *Choice in Welfare*, etc. Secretary of Mont Pelerin Society and Wincott Foundation. Member of Political Economy Club. Council Member of University College, Buckingham. Lectures and writes widely on post-war policies and economic requirements of free society.

E. VICTOR MORGAN. Born 1915, educated at Warwick School and Sidney Sussex College, Cambridge; post-graduate student, Christ's College, Cambridge, 1938-41; Professor of Economics, University College, Swansea, 1945; Professor of Economics, University of Manchester, 1966-74; Professor of Economics, University of Reading, from Jan. 1975. Author of *The Theory and Practice of Central Banking, 1797-1913* (1943); *Conquest of Unemployment* (1947); *Studies in British Financial Policy, 1914-25* (1951); *A First Approach to Economics* (1955, 3rd edn. 1972); *The Structure of Property Ownership in Great Britain* (1959); (with W. A. Thomas) *The Stock Exchange* (1964, 2nd edn. 1969); *A History of Money* (1965); *The Economics of Public Policy* (1972). The IEA published his *Monetary Policy for Stable Growth* (Hobart Paper 27, 1964, 3rd edn. 1969). IEA Advisory Council.

ALAN ARTHUR WALTERS. Born 1926, educated at Alderman Newton's Secondary School, Leicester, and University College, Leicester (now the University of Leicester): BSc (Econ.) (London) with First Class Honours, 1951; research student at Nuffield College, Oxford; Lecturer in Econometrics, University of Birmingham, 1952-61, then

[14]

Professor and Head of the Department of Econometrics and Social Statistics, 1961-68; Sir Ernest Cassel Professor of Economics, University of London, at the London School of Economics, since 1968. Member of the Commission on the Third London Airport (Roskill Commission), 1968-70. Publications include: *The Economics of Road User Charges* (1968); *Introduction to Econometrics* (1969, 2nd edn. 1971); (with E. Bennathan) *The Economics of Ocean Freight Rates* (1969). Author of Hobart Paper 44, *Money in Boom and Slump* (1969, 3rd edn. 1971). IEA Advisory Council.

MICHAEL PARKIN. Born 1939, educated at the University of Leicester: First Class Honours, 1963. Professor of Economics, University of Manchester, since 1970. Has taught at the Universities of Sheffield, Leicester and Essex and held visiting Professorships in the United States, Australia and India. Writes, lectures and broadcasts on inflation and related topics. His research has yielded papers assessing the effects of incomes policies in *Incomes Policies and Inflation* (1972), edited jointly with Michael T. Sumner. Editor of the proceedings of the Conference of the Association of University Teachers of Economics and of *The Manchester School*.

SAMUEL BRITTAN. Born 1933, educated at Kilburn Grammar School and Jesus College, Cambridge (First-Class Honours in economics); various posts on the *Financial Times*, 1955-61; Economics Editor of the *Observer*, 1961-64; an Adviser at the Department of Economic Affairs, 1965; principal economic commentator on the *Financial Times* since 1966. First winner of the Senior Wincott Foundation Award for financial journalists, 1971. Research Fellow of Nuffield College, Oxford, 1973-4, Visiting Fellow since 1974. His publications include: *Steering the Economy* (third edition, 1971); *Left or Right: The Bogus Dilemma* (1968); *The Price of Economic Freedom: A Guide to Flexible Rates* (1970); *Is There An Economic Consensus?* (1973), and *Capitalism and the Permissive Society* (1973). For the IEA he has written *Government and the Market Economy* (Hobart Paperback No. 2, 1971).

HARRY GORDON JOHNSON. Born 1923, educated at the Universities of Toronto, Cambridge and Harvard. Various teaching posts, 1943-49; Lecturer in Economics at Cambridge and Fellow of King's College, 1950-56; Professor of Economic Theory at Manchester

[15]

University, 1956-59; Professor of Economics, University of Chicago, since 1959; Professor of Economics at the LSE (succeeding Lord Robbins), 1966-74; President of the Canadian Political Science Association, 1965-66. His published contributions to economic literature, particularly in monetary theory and international trade, have been counted at over 200. His most recent books include: *Essays in Monetary Economics* (1967); *Aspects of the Theory of Tariffs* (1971); *Macro-economics and Monetary Theory* (1972); *Inflation and the Monetarist Controversy* (1972); *Further Essays in Monetary Economics* (1973); etc. Joint author (with John E. Nash) of Hobart Paper 46, *UK and Floating Exchanges* (1969). IEA Advisory Council.

The Permissive Economy
SIR JOHN HICKS

Drummond Professor of Political Economy,
University of Oxford, 1952-65
Nobel Laureate, 1972

THE FIRST thing that needs to be said about our present troubles
is that they are not of a monetary character, and are not to be cured
by monetary means. Of course it was true, in the old days, that
inflation was a monetary matter; prices rose because the supply of
money was greater than the demand for it. That might happen, as
in the 16th century, by increased supply of monetary metals; or
it might happen, as in many later instances, because paper money,
or abstract bank money, was unduly increased by mistaken policy,
or for some public purpose to which monetary stability was regarded
as secondary. But the age when these things were true is now over.
Money is now a mere counter, which is supplied by the banking
system (or by the government through the banking system) just as
it is required. This has been the fact, at least since 1945. The
attempts which have occasionally been made to reimpose some kind
of monetary control of the old type have done no more than demon-
strate their own futility.

Budgetary control of inflation weak
It was widely supposed that monetary control could be dispensed
with, because a substitute for it had been found. The responsibility
for avoiding (or moderating) inflation had been transferred from the
banking system to the Government's budget. But though the
objective of budgetary policy was set as 'full employment without
inflation', it soon became apparent that government spending (and
taxation) had a much more direct effect upon employment, and 'real'
activity generally, than it had upon prices. Governments, indeed,
having taken responsibility for employment, were more sensitive to
the effects of their actions upon employment than on the value of
money; but more than that was involved. It is technically easier for

[17]

a budget to be expansionary, or contractionary, in its effects on employment, than it is for it to operate, at all strongly, on the value of money.

If money wages are fairly steady and productivity is increasing, prices are likely to fall; if money wages increase no faster than productivity, prices can be steady. But there is nothing in the system which prevents wages rising faster than productivity; nothing therefore which automatically stops, or controls, inflation.

Floating exchange rate weakens constraints on wages
How was it then that in the 'fifties and 'sixties, when we already had this kind of economy – a Permissive Economy, we may call it – inflation was not more severe? There were, I think, two reasons. One was the importance that was still attached, up to 1967, to a stable rate of foreign exchange. So long as the rate of exchange was regarded as sacred, wages in export industries, and in industries that competed with imports, could not move freely. Attempts might be made to shake off the constraints by pressure for protection, but the international movement in favour of freer trade kept that in check. So there were some wages, some important wages, which were under some pressure from external sources – at least to the extent that they could not move far out of line with corresponding wages in other countries – and that, in its turn, imposed some constraint upon wages generally. But then, with the floating of currencies, that constraint was removed.

There was however another reason for the fairly moderate behaviour of wages in the 'fifties and 'sixties. There was a hangover from the days when trade unions thought of themselves as primarily defensive organisations. Their object was the maintenance of what they understood to be *fair* wages, 'fair' (it is true) in several senses, not always consistent with one another. One aspect of fairness is the maintenance of customary differentials; another is steadiness over time; another is fairness in relation to the profits of the employer – if the employer is earning big profits it is fair that his employees should get their share of them. These principles are not necessarily de-stabilising, not (at least) very de-stabilising; but as time went on, they began to take more dangerous forms. War and post-war experience already attracted attention to the relation of wages to prices; so the pursuit of steadiness over time became a pursuit of steadiness in real wages. Then, with experience of rising real wages (such as in this period were indeed experienced) it became a pursuit of rising real wages at a customary rate. Meanwhile, the conversion of several large industries into state monopolies transformed the old relation between wage claims and profitability; for when the state was the employer, there was nothing that this employer could not (apparently) be made to pay. But it was only by degrees that the fact that these fences were going down became at all clear. During the long boom of the 'fifties and 'sixties they still, after a fashion, held.

There was always a chance that they would hold, at least for a time; so I do not believe that I was wrong, in the first analysis which I gave of what I am here calling the Permissive Economy,[1] in concluding that what I was saying was 'rather optimistic'. It was, after all, to give us about 20 years of fair prosperity; and that is not a bad showing. I was nevertheless very careful (in the same paper) to emphasise that the stability of such an economy was very fragile; so that any serious disturbance to the rate of progress that was attainable would be very dangerous. That is what has happened.

The forms of unemployment

This is no place to discuss the causes of the primary product crisis (for it is much more than an oil crisis) that has come upon us since 1972-73. It is enough to say that it marked a sharp fall in the real earning-power of British labour, in terms of the things that British people want to buy. And now that the veil of money has been so largely stripped off, we have to face it in terms of traditional economics, waking up from the sweet Keynesian dreams that have been with us for so long. When the real demand for labour falls (as it has fallen) then, traditional theory says, either there must be a fall in real wages, or there must be unemployment. That is true, profoundly true; but we must be careful how we interpret it. There is a good deal of pretence about unemployment; there are already many, who are unemployed in the sense of this proposition, but whom we do not reckon as such. Some are non-employed, who by some social arrangement or other have been withdrawn from the labour market.

[1] In a paper entitled 'Economic Foundations of Wage Policy' which I gave to the British Association in 1955. It is reprinted in my *Essays in World Economics*, Oxford University Press, 1959; see especially p. 104.

Others are nominally employed, their employer having been prevented from declaring them redundant, though by their retention they hamper the future expansion of those firms, and may even make 'rescue operations' necessary to prevent closing down.

It is clear that the British economy, in real terms, is in a slump. We had already learned that slump conditions can co-exist with rising prices; we are now learning that a slump is consistent with apparently high employment, since employment can be 'cooked'.

Can real wages be reduced?

Pseudo-employment, like unemployment, is an evil; but it is no cure for the one, either one, to induce the other. What is needed is a clear recognition of the plight that we are in; so that, having faced the worst, we can turn our energies to the organisation of recovery, sound recovery. This means an acceptance of the fact that the real wages of labour in general have got to be reduced – wages in the widest sense, including the social services that are part of the cost of labour. They have got to be reduced until we can bring about such a rise in real earning-power as will enable the rise in wages to be resumed.

That, obviously, is hard enough; but it is made much harder by the fact that the fall cannot, or should not, be universal. There has been a general fall in the earning-power of British labour; but it has come about in a particular way, so that the fall is not uniform. Those who can produce at home the things (or substitutes for the things) that have become harder to get from abroad are more highly needed than they were. That is why it was so foolish to let the political crisis erupt over miners' wages, since miners' wages, in a crisis like the present, ought to rise relatively to the wages of others. We cannot, in this sort of emergency, do without miners; but there are not many others who have a similar claim to privileged treatment.

We shall not emerge from the slump that we are in until there has been an accepted fall in the real wages of most sorts of labour (which must no doubt take the form of a rise in prices which money wages, and of course salaries, make no pretence to catch up). It is the attempt to bring about such a fall in real wages by a lag in money wages behind money prices – a lag which has by now become quite undependable – that is the principal cause of the acceleration of inflation.

[20]

Realities, not gadgets

Economists, we are told, have no solution to the present emergency; if what is meant is that they have no gadget to deal with it, that is true. This is not a time for gadgets; it is a time for facing realities. Britain, like other industrialised countries, has taken a blow; but it is a blow from which we can recover. But we shall not begin to recover until we recognise where we are, and from what point the recovery has to begin. To pretend that we can carry on in the old ways, as if nothing had happened, merely prolongs the agony.

The Dispersive Revolution

E. H. PHELPS BROWN

Professor of the Economics of Labour,
University of London, 1947-68

THE ONRUSH of inflation is due in great part to mounting pay claims. These claims are enforced by the increased bargaining power of groups of all kinds of employees, and their increased readiness to use that power independently. The 'shift of power to the shop floor' is only one manifestation of the dispersion of power from the centre to the base, from accepted institutions and procedures to direct action by the employee. This dispersion has come about in a number of Western countries in recent years. Its implications for the working of their economies constitute a revolution.

Its basic cause is a two-fold change in the attitude and expectations of the employee – a quarter of a century's experience of full employment has warranted the assumption that he can push up pay and strike without risk of losing his job; and the experience during the same years of a progressively rising standard of living has fostered the expectation that the standard will continue to rise year after year. This expectation has been heightened by the example of some outstanding gains, widely cited without reference to all their conditions, in particular the period over which they extend. When the cost of living rises, the real gain that had been supposedly afforded by the last settlement is whittled away, and the next claim appears essential to defend the standard of living. Those who had no initial wish to press their claims cannot afford to be left behind. Readiness to combine and strike has spread among types of employees where not long ago it was unknown.

Rise of the power of the strike

At the same time, for a variety of reasons, the power of the strike has increased. Some industries have become more concentrated, so that the stoppage of a few firms will cut off a large part of the output. Industry has also become more integrated, so that withholding a

[22]

particular supply, it may be by only a small number of employees in one place, stops production over a wide area. Often the function of the strike is not simply to withhold labour from the employer pending agreement on the terms of employment but to inflict disruption on the community. The employer is now the salaried manager, more concerned to keep the business going than to maintain profits in the immediately foreseeable future; the firms under the severest financial constraints may be the least able to 'take' a strike.

The dispersion of an economic power that is at the same time much increased leaves the Western economies without institutions or principles for controlling the movements of pay. Our established and well-tried machinery of industrial relations now works to very different effect or is by-passed. The trade union itself is changed radically: its officers are no longer able to enter into agreements which they then hold their members bound by, but must submit proposed terms to them for approval or rejection, and have to tolerate action taken by their members locally without reference to them. Negotiations at the national level consequently exert much less control over the movement of pay locally.

The powers of government are likewise reduced. Where the trade union was a single body guided by its national officers, governments could exert an effective influence by gaining the co-operation of those officers, and might rely on statutory powers in the last resort; but where the members of the union are determined for their own part, individually and collectively, to stay away from work, no government in a free country can compel them to go back.

The working of our institutions has changed faster than our ideas about them. The present use of the right to combine and to strike continues to be legitimised by the place that right held in an earlier and different setting. Negotiations are still seen as being for the division of the product between employees and employers, when in practice they divide only a total available to employees as a whole between different groups of them according as they push ahead or lag behind in raising their pay. After more than one demonstration of the inability of governments armed by Parliament with statutory powers to hold back the pay of a group of employees who control a vital supply, governments are still held responsible for checking inflation, and there are still advocates of a statutory incomes policy.

[23]

Generally, we have tried to regulate the movement of money incomes by reviving the procedure that served us so well during the Second World War, namely by working through institutions for collective bargaining at the national level, according to an understanding between the government and the national leaders of the trade unions and employers' associations, an understanding that was assured of at least the passive assent of the rank and file. But now the rank and file have ideas, and power, of their own. Policies can be effective only in so far as the need for them is accepted by employees in their millions.

Passive control via the money supply?

If the need is for a widespread change in individual attitudes and expectations, does not control of the quantity of money provide just the pervasive influence required to bring that about? Only the progressive expansion of that quantity has made possible the continued pushing up of labour costs in past years without loss of jobs, an experience that has fostered the attitudes and expectations which carry the inflation on. Can a more realistic outlook be restored unless the expansion is checked?

Let us be clear about how the checking will take effect. As it meets the overrun of commitments recently entered into, it will bring a crisis of liquidity. In the longer run most firms and their employees will find that higher labour costs can be covered by higher prices only under penalty of lower sales and fewer jobs. There can be no question of the generality and penetration of this constraint, or of its ultimate inescapability. But it will have been imposed, visibly, as a measure of deliberate policy, and perceived as such by employees. The question is, how will they react as electors?

Their outlook might be changed radically if monetary constraint coincided with other measures, such as rationing of food or fuel, that an economic crisis demanded, and that drove home the lesson that times have changed. But short of such an experience, what grounds are there to expect that the electorate would allow the continuance of bankruptcies and unemployment that had been imposed as an act of policy and that a policy of reflation could remove? Were widespread understanding of the need to accept monetary discipline attainable, could we not equally well attain the acceptance of discipline in pay claims that would make monetary pains and penalties unnecessary?

Easy street or common cause?

If we are so far from such acceptance now, it is not because those who are pressing their claims are more selfish or less reasonable than those of us whom the claims appal. If we ask them why, when our balance of payments and our inflation are what they are, they persist in action that can only make both worse, we shall get two kinds of answer. One will show no understanding of the danger, or will regard it as remote, as only one more scare that will pass as previous crises of sterling have done, leaving us still in easy street. The other kind of answer accepts the reality of the danger and the need for common action, but insists that the action be indeed common: meanwhile prices go up and up, others are getting big pay rises, and what good will it do to the country compared with the harm it will do to me if I myself hold back? There are two basic needs here – for understanding, and for 'fairness'.

A diffusion of understanding urgently required

Our lack of understanding may come to be seen as the main cause of our inability to respond to the need of the hour. In 1940 that need was made unmistakable by the guns of the enemy at Calais and his bombers overhead. The economic needs of today can be made unmistakable only by plain speaking. But of this there has been an amazing lack. Perhaps politicians are reluctant to tell voters, most of whom are employees or the wives of employees, that they should not press the claims they see as vital. Perhaps no one likes to call a halt on those who are asking for much less than he himself is getting already.

But the dispersion of power calls for a diffusion of understanding. Not until the senselessness and the perils of present procedures are clear to those who have the power to press their claims will they be prepared to consider changing what they see now as the indispensable defences of their own standard of living. It is hard to see how a way forward can be found, in a free country, save by some mutual agreement to give up the war of all against all – by a 'Social Contract' in the original sense of that term. But the widespread willingness to entrust the movement of one's own pay to new procedures depends

[25]

on no less widespread appreciation of where the present ones are taking us.

Criteria of 'fair comparison'

It depends also on the new procedures being grounded in criteria of 'fairness' accepted by those concerned. When the power to raise pay is dispersed, the one basis of order is the conviction of the majority of employees that pay should be 'fair', and that they are warranted in pressing their own claims only in proportion as they can show them to be justified. Because the great ground of justification is 'fair comparison', the ordering of the rise of pay depends on the acceptance of relativities. Last January's report on *Relativities*[1] should be appreciated not as the swan song of the Pay Board but as the opening of a prospect of great constructive potentiality.

In the shorter run, judgements about what pay rises are 'fair' depend much upon the cost of living. It does not seem 'fair' to the employee that prices should go on rising through the months in which his own pay stays the same. Our recent threshold agreements have proved expensive because of changes in world prices unforeseen when they were drawn up, but it does not follow that the aggregate rise in pay by mid-1975 will have been larger than it would have been without them. When negotiations are opened in the knowledge that the rise in the cost of living since the last settlement has already been offset, and that whatever agreement is reached will provide for an automatic continuance of this off-setting, what grounds are there for a claim? Grounds there may be, but they will have had to be narrowed and clarified. Indexation is 'inbuilt inflation' only on the assumption that this separation of issues does not result in a lower rise in all than when a single inclusive figure is envisaged.

Lessons from Sweden

In these various respects we can learn from the recent example of Sweden.[2] The relative ease with which comprehensive settlements were reached there in the first months of 1974 has been attributed to the previous working out of adjustments and the adoption of provisions that met the Swedish employee's strong sense of the need

[1] Pay Board, *Advisory Report 2*, Cmnd. 5535, HMSO, January 1974.

[2] S. Nycander, 'Collective Wage Negotiations in Sweden', *Skandinaviska Enskilda Banken Quarterly Review*, 3, 1974.

for 'fairness' in the pay structure. White-collar pay had been brought into a widely acceptable relation with manual. Among the manual workers, pay under time-rates was already being raised automatically to match the upward drift of earnings under piece-rates, and on the motion of the trade unionists themselves the lower paid had been raised relatively to the higher. When the parties entered the negotiations of last winter they were thus able to consider what the economy could afford without distraction by the need of particular groups to catch up the others, or fear of the chain reactions that would ensue.

Our own new procedures have to grow up in our own soil, but the Swedes show us at least that some measures that meet the needs of the time are not inherently impracticable.

An Effective Social Contract

JAMES E. MEADE

Professor of Political Economy,
University of Cambridge, 1957-68

THERE IS now every prospect of a major economic crisis in the near future. The existing so-called 'Social Contract' is proving quite inadequate to prevent a rapid inflationary rise in costs and so in prices. The inflationary spiral shows signs of exploding rather than slowing down. If this occurs, at some point there will be a run on the pound as holders of sterling come to realise the imminent collapse of the currency. We have at present a huge current balance-of-payments deficit and a large overhang of debts already incurred to the foreigner. The result for a country like the United Kingdom, which is not only incapable of repaying debt but must for some considerable time by hook or by crook continue to borrow annually large sums from foreigners in order to finance essential imports of foodstuffs and raw materials, will be catastrophic.

Full employment, free bargaining

The moment of truth is rapidly approaching when we shall no longer be able to continue on our present course. At present we run our affairs with a set of principles and institutions which might well have been expressly designed to lead to this catastrophic result. On the one hand, our monetary and fiscal policies are conducted on the general acceptance of the need to maintain the total flow of monetary expenditures on goods and services at a level which will maintain full employment of labour regardless of what happens to money, wages and prices. On the other hand, we simultaneously maintain that there is one set of powerful monopolies (namely, the labour unions) each one of which must remain completely free to exercise its monopoly power in order to settle its own money rate of pay. Human nature being what it is, each group uses its monopolistic power with the aim of getting a somewhat bigger rise than the average or at the least as big a rise as the average. This attempt to

[28]

get a quart out of a pint pot of available goods leads to an explosive inflation of money prices and costs; and the introduction of threshold payments or similar arrangements then adds a guarantee that the pint pot will provide the quart which is demanded. Economists can be justly criticised for many errors of commission and omission in the advice which they have given; but it is misguided to hold them up to ridicule (as is so often done) for not knowing how to get the quart out of the pint pot.

In order to put an end to this foolish way of running our economic affairs it is necessary to operate simultaneously on three fronts.

(1) We must curb the rate of growth of total money expenditures on goods and services.

(2) In order to prevent this from leading to unemployment we must find some way of ensuring that money wage-rates are not pushed up at a speed which outruns the moderated and curbed rate of growth of money expenditures on goods and services.

(3) We must shift the emphasis away from wage settlements on to fiscal and other measures for ensuring a just and socially acceptable distribution of income and wealth.

Social control

I shall outline below a set of arrangements which in my opinion would achieve these three objectives in an acceptable manner. But before doing so I must try to forestall one criticism. My proposals are not, repeat not, an exercise in trade-union bashing. They do, however, face the fact that labour unions can be extremely powerful monopolistic organisations and that one cannot run an economy unless powerful monopolies are subject to some elements of social constraint. To say that a monopoly should accept some degree of social control is not the same thing as saying that it should be destroyed or crippled. To require ICI to observe certain rules in setting its selling prices is not to cripple, destroy, or otherwise bash ICI.

Indeed, I would assert that the acceptance of proposals of the general kind which I shall outline is a necessary condition for the continued existence of trade unions as effective independent bodies in a free, democratic society. The present system will not in practice be allowed to continue indefinitely. The basic danger is that when the collapse comes power will be taken by an authoritarian régime on the right or the left of the political spectrum. Recalcitrant and independently-minded trade unionists will be among the first to be dosed with castor oil, sent to Siberia, put into a lunatic asylum, shot in the back of the neck, or treated to whatever may then be the fashionable technique of the ruling fascist or communist régime. The proposals which I shall outline are based on the assumption that we can achieve a fair consensus of opinion (including the opinions of all moderate trade unionists) that some degree of social restraint over wage setting is a necessary condition for the continuation of an independent trade union movement in a free society.

An outline proposal
The following is a very brief sketch of one set of proposals which might meet the situation.

(1) *Growth path for pay*
Something must be done to restrain the growth of total money demand. I propose that domestic monetary and fiscal policies should be expressly devised and ruthlessly applied to control the total level of money expenditure on goods and services on the following lines. A growth path for the total amount paid out in wages and salaries in the community would be set at, let us say, 20 per cent over the coming year, 15 per cent over the following year, 10 per cent over the third year, and 5 per cent over each succeeding year. If total wage and salary payments fell below this growth path, total monetary demand would be stimulated; if they rose above it, demand would be restrained.

If necessary, special steps would be taken to improve and to collect more rapidly statistics of total wage and salary payments and to enable the Chancellor of the Exchequer and the Governor of the Bank of England together to take prompt and effective tax and/or monetary measures to inject or to withdraw money purchasing power so as to keep the index of total wage and salary payments on course.

If this were done, the degree of unemployment (apart from frictional and structural unemployment which need quite different treatment) would depend upon the fixing of money rates of pay. If monetary and fiscal policies were successfully devised to cause the total amount paid out in wages and salaries to rise by 20 per cent in the first year, then the level of unemployment would rise or fall according as the average level of pay per employee was raised by less or by more than 20 per cent. Any group which demanded and achieved a higher rate of increase than 20 per cent would imply that some other group must receive a less than 20 per cent rise or that the numbers in employment should fall.

But the system must not be allowed to lead to unemployment, which it would most probably do if no changes were made in present wage-fixing arrangements. Many monopolistic groups might insist on getting more than a 20 per cent rise and most others would insist on at least a 20 per cent rise with the result that on the average earnings per head were pushed up by more than 20 per cent, in which case unemployment would result.

(2) 'Excessive' pay claims

It is reasonable to hope that, with a system of this kind (together with assurances about profit margins and so prices which I discuss below), the trade union movement as a whole would voluntarily exercise restraint in any demands for increases in pay above the 20 per cent level. But it is virtually certain that there would be some unreasonable claims and that some further restraining measures would be necessary. Indeed the TUC cannot control the constituent trade unions and the large trade unions cannot control the shop-floor action of their members. Some social restraint over individual pay settlements would, therefore, be essential if unemployment is to be avoided. For this purpose it is necessary to choose between an incomes policy which regulates the whole structure of pay with sanctions for breach of the rules or, alternatively, a system which leaves the settlement of individual rates of pay to free collective bargaining, but with some restraint over the bargaining power of those who demand *excessive* rises in pay. My preference would be for the latter course which might take the following form.

All employers and workers would remain free to settle for any wage bargain they could agree upon. There would, however, be laid down a 'norm' for increases in pay per head (taking into account fringe benefits) which was slightly below the average rise in total earnings that was being maintained by the monetary and fiscal control of total demand. Thus if the total amount paid out in wages and salaries were planned to rise by 20, 15, 10, and 5 per cent over the first, second, third, and subsequent years of the new system, the 'norms' for rises in pay per head might be set at 18, 13, 8, and 3 per cent respectively. If any employer was ready to agree to pay increases in excess of the 'norm' (for example, in order to attract labour), he would be perfectly free to do so; and there would in practice be many such settlements. But if any employer wished to resist a claim that was in excess of the 'norm' (for example, because the market for his product would not stand the increase in costs), there would be a set of independent tribunals to determine whether the claim were indeed in excess of the current 'norm'. The workers concerned would in any case be free to take industrial action to press their claim. If their claim was not in excess of the norm, their bargaining power would remain unimpaired as at present. But if they took industrial action to press a claim which had been judged to be in excess of the 'norm', they would be subject to certain financial disabilities: they would lose any accumulated rights to redundancy pay; any supplementary benefit paid to their families would become a debt which they would have subsequently to repay; and any strike pay which they received would be subject to a 50 per cent rate of tax.

(3) *Restraint of profit margins by price controls*
These proposals are designed only to restrain excessive rises in pay. Such a system would be totally unacceptable if it were not accompanied by measures which restrained the profit margins earned in industry. For this purpose, in addition to the statutory prohibition of various monopolistic restrictive practices, reliance should be put upon price control arrangements which prevented the setting of profit margins above the levels necessary to earn an adequate return on capital.

There can be no absolute guarantee about movements of the cost of living. But if productivity per worker increased at 2 per cent a year and if profit margins were effectively controlled, average real earnings per head would rise by 2 per cent a year unless either of two things occurred. First, the terms of international trade might change. If they improved, real earnings per head would go up by more than 2 per cent; if they worsened, real earnings per head would rise by less than 2 per cent. But that would be the expression of the inevitable fact that uncontrollable overseas events had occurred to damp down the rate of improvement in the country's standard of living. Secondly, the government might need more or less tax

[32]

revenue for public purposes and it might decide to obtain it by raising or lowering the rate of indirect taxation (e.g. VAT). If so, real earnings per head would, of course, be affected through the consequential rise or fall in the taxed price of goods and services. But that would be the express purpose of the tax change.

(4) *Redistribution*

Finally, for a just and socially acceptable distribution of income and wealth the emphasis must be put upon the taxation of rich and wealthy individuals and upon aid (by social benefits or otherwise) to the poor and less prosperous members of society. The control of profit margins and the taxation of *companies* should be such as to allow efficient firms to make really adequate profits on their capital. Redistribution should be achieved by the taxation of the incomes and properties of wealthy *individuals* and the assistance of less prosperous families and individuals.

Far-reaching egalitarian measures are possible, and in my view desirable, on this basis of redistribution between individuals, provided that the complex spiral of wage-cost-price-income inflation can be checked by the kind of method outlined here. This indeed is the basis for a true and effective social contract for execution by a social-democratic government.

Oil and The Crisis
RICHARD KAHN
Professor of Economics,
University of Cambridge, 1951-72

IN THE first ten months of 1974 the exports of West Germany to
the oil-producing countries amounted to DM 7,900 million, about
£1,200 million, an increase of DM 3,000 million, about £450 million,
or 61 per cent, over those of the same period of 1973. In the first
eleven months of 1974 the exports of the United Kingdom to the
Middle East amounted to £1,100 million, an increase of £350
million, or 46 per cent, over those of the same period of 1973.

(This essay is written abroad. The material available is not sufficient for a precise comparison between the UK and Germany. A
useful background against which the figures can be appreciated is
the prospective magnitude of the 1975 total exports of the UK of
about £15,000 million.)

Of the German exports in 1974 to the oil-producing countries,
one-third went to Iran, worth about £400 million, an increase of
one-half over 1973. UK exports to Iran at £240 million increased
by 60 per cent, but even then they amounted to only one-fifth of
total UK exports to the Middle East, and to little more than one-
half of those of Germany.

The reason for emphasising exports to Iran is that an exceptionally
high proportion of the increase in her oil profits is already being used
to pay for additional imports.

Germany's oil-saving efforts
In the first ten months of 1974 Germany's expenditure on imports
of oil was DM 19,000 million, about £2,800 million, an increase of
DM12,000 million, about £1,800 million, over the same period of
1973, in spite of the volume of imports of oil being 8 per cent lower.
The dramatic rise in the price of oil took place towards the end of
1973. Thus one-quarter of the German deficit caused by the rise in the
price of oil was recouped by additional exports to the oil producers.

[34]

In the first nine months of 1974 the expenditure of the UK on imports of oil was £4,150 million (seasonally adjusted), an increase of £3,000 million over the same period of 1973, the volume of imports of oil being only 3½ per cent lower. The increase in the exports of the UK to the Middle East in the first eleven months of 1974 over the same period of 1973 contributed only £350 million.

On the assumption that the sterling prices of UK exports to the Middle East rose between 1973 and 1974 by the same amount as those of all UK exports on the average, i.e. by 28 per cent (in the first nine months), the 46 per cent increase in the value of UK exports to the Middle East in the first eleven months of 1974 amounted to only 14 per cent by volume. The 60 per cent increase in the value of exports to Iran amounted to only 25 per cent by volume.

French exports to Iran
The French Prime Minister, M. Jacques Chirac, after a recent three-day visit to Iran, was able to report the signature of several very important contracts for a total of about $6,000 million – about £2,500 million – on projects which had been under discussion since early in 1974. They include the construction of a 40-mile subway system in Teheran, a 'special steel' mill, and a factory for the production annually of 100,000 Renault cars, building up to 200,000 at a later date. A French company is to set up a colour television system. There is to be a joint venture for transporting liquified natural gas. Provision has been made for the manufacture of chemicals in Iran, for developing telecommunications, and for the sale to Iran of railway equipment. 200,000 housing units are to be constructed by French contractors over eight years, together with the ancillary hospitals and schools.

UK's record non-oil deficit
If orders of such magnitude could be secured for Britain, her exports to Iran alone would transform her balance-of-payments position. Already in November, France had achieved a balance of visible trade, including the whole effect of the rise in the price of oil. The UK visible balance was adverse to the extent of £534 million (seasonally adjusted on a balance-of-payments basis), of which £307 million was the deficit on trade in oil (including about £50 million for the value of imports of oil at the old price). Thus, with the omission

of oil, the visible deficit in November was £227 million – back to the record level at which it stood at the end of 1973.

It is clearly unsatisfactory to base an appreciation of the prospects of UK exports to the oil-producers on vague statements, but in the absence of hard facts it is impossible not to feel deeply pessimistic, while the figures for 1974, quoted earlier, comparing the performance of the UK with that of West Germany, are not encouraging.

One does hear a considerable amount of talk phrased in such terms as 'possibilities are being explored', 'missions are being proposed', and 'scientific and technical co-operation'. I have been told of an inquiry from an oil-producing country for a fairly modest amount of the bread-and-butter product of a large British producer suffering from a serious fall in demand – a product which takes only a few months to manufacture. A delivery period of 18 months was quoted. The order for an equivalent product was placed in another country.

I have been told that the representative of a large British firm, when he had had emphasised to him the importance of providing servicing and spare-part facilities, retorted: 'What! Would you have us lock up capital in a country like Iran?'

'An astonishing delusion'
We are suffering from an astonishing delusion about the adverse effect of the rise in the price of oil on our balance of payments. It appears to be based on the assumption that the profits on oil will remain available to finance the deficit caused by the rise in the price of oil; and that it is simply a matter of securing our share of that finance.

To secure our share of the *available* finance is not all that difficult. London offers great attractions to the oil-producers, in spite of some lack of confidence in sterling. New York is our main rival. But London banks are rightly regarded as sound – in a world in which confidence in financial institutions is suffering severe shocks. In Zurich newly-placed foreign money is taxed. Neither Frankfurt nor Paris offers much attraction and Tokyo is an unwelcoming host to foreign funds. It is true that Euro-currencies, other than Euro-sterling, can be held with London banks. But sterling held in London offers higher rates of interest, which are the offset to some lack of confidence in sterling.

One has the impression that far more effort is being devoted to

[36]

securing the investment of oil profits in this country than in exporting our products in exchange for oil profits. Our country is fast becoming a land of financial borrowers rather than of industrial exporters.

The profits of the oil producers are being rapidly devoted to paying for additional imports. It is tremendously more important that our industrialists should be successful rather than our financiers. To the extent that failure on the part of our industrialists forces us to rely on our financiers, the law of compound interest will relentlessly lead to our becoming the insolvent subject of the oil producers.

The UK needs not merely to secure her proper share of the additional trade with the oil-producing countries. She needs to do much better than that. For her balance of payments was already in bad shape before the price of oil was raised.

Iran, Libya, Algeria and Nigeria, in particular, are placing huge orders; and those which are being placed by Saudi Arabia are also large, even though they represent a relatively small proportion of her profits, which are enormous. The imports, in recent months, of the oil-producing countries provide no indication whatever of the dimensions which their imports will assume in a few years' time, for there is usually a considerable interval between placing an order and the passage through the customs of the resulting export.

UK industrialists' failure
There is every sign that our industrialists are letting us down very badly indeed. In October and November the volume of our exports was running 8 per cent below the recent peak. The excuse which is advanced is the world-wide cutback in industrial production. But the volume of international trade in manufactured goods is, I believe, still rising, though slowly – the result, I suggest, of the increasing imports of the oil-producing countries.

Between January and September, 1974, the monthly placing of new export orders with our engineering industries fell by 17 per cent, with the result that the total size of the export order book had begun to fall in September. In this context the *Financial Times* of 27 December, 1974, referred to 'less buoyant economic prospects in the UK's main overseas markets following the oil crisis'. But the oil crisis also means surplus capacity in our export industries and *buoyant* markets in the oil-producing countries.

[37]

While our balance of payments is deteriorating, most of the industrial countries which were experiencing balance-of-payments trouble – especially France, Japan and the United States – are improving their positions. The hideous fact is that, on present form, the UK is one of the least successful industrial countries – almost certainly *the* least successful large industrial country – in securing the orders which the oil-producing countries are placing at a rapidly increasing rate.

Summary: drastic changes in economic policy required
If I had to sum up in short the position in 1975 I would say:

1. Even if other things are not equal, Britain will – and is not merely likely to – in 1975 confront a worsening crisis that will call for changes in government policy – not merely significant but drastic.

2. The principal reasons for the present economic crisis are:
(i) the rise in the price of oil;
(ii) the failure of British industrialists to secure an adequate share of the orders by the oil producers financed by their profits;
(iii) trade union pressure;
(iv) the abandonment by the Government of incomes policy.

3. In 1975 the crisis will take the form of a worsening balance-of-payments position and an increased rate of inflation.

4. The result of leaving present Government policies unchanged would be disastrous.

5. A drastic change of economic policy is called for. Apart from re-introducing an effective incomes policy, designed to prevent the country's competitive position from being further eroded, the Government should avoid providing industrialists with excuses – no matter how feeble – for their poverty of performance. While in the longer run there is much to be said for Wedgerry-Bennery – especially in so far as it is calculated to raise the lamentably low level of British industrial management – it is quite irrelevant in the context of present dangers.

Industrial Investment and
Interest Rates

HENRY SMITH

Vice-Principal, Ruskin College, Oxford, 1947-70

THE CENTRAL thesis here submitted is that the maintenance of an expanding free enterprise sector in the world economy is likely to demand a long-term decline in interest rates. This argument has two foundations. The first is that as capital has accumulated so fast in the last 20 years, the relative abundance of work-places and the high cost of providing them, combined with the high wages and increasing personal resources of labour and increasing public provision for poverty, have created a position in which the natural bargaining power of labour has increased.

Keynes, in his enigmatic last chapter,[1] speaks of the impending decline in the return to the functionless ownership of capital, but does not spell out to whose obvious advantage this must be. Any recovery of the free enterprise sector must be based on the free play of bargaining power as it emerges in the course of economic development. This, in effect, means the accumulation and effective utilisation of capital. This argument is general in its application: it presents no difficulties which a society of reasonable men cannot overcome – but also no guarantee of automatic recovery without a lot of common-sense and goodwill.

Decreasing returns in agriculture and extraction

The second reason why it looks as if a decline in the marginal productivity of capital must be recognised as inevitable is specific to our times. The development and application of medical science has led in the last 30 years to an unprecedented growth in the rate of increase of world population and, in developed countries, in the expectation of life. This has in turn led to rising relative food prices, and to an increasing burden of support for the elderly and economically inactive. The increasing support is inflationary in the simplest

[1] *The General Theory of Employment, Interest and Money*, Macmillan, 1936.

[39]

sense of the word: it adds to money income without increasing the supply of consumer goods, and the result is the same whether provision is private or made by the state. The rise in relative food prices is inflationary because it puts up the cost of living and stimulates wage claims. Superimposed on all this, the increasing dependence of our main industries on the use of mineral fuel and upon the production of goods which consume mineral fuel has put the owners of mineral resources in a position of economic power which monopolistic organisation has enhanced. In the phraseology of classical economics, the free enterprise sector has run into a patch of decreasing returns, in agriculture and extraction.

If inflation is regarded as an excess of demand over supply for consumer goods, and if we have run into a phase where supply is inelastic due to the rising import cost of food and fuel, and if money incomes are hard to reduce because of union power and a built-in resistance to tax increases (partly also due to union power), the immediate need is to increase the flow of output from increasing plant and labour supplies. Any increase in the supply of consumer goods which exceeds, however marginally, the increase in demand for consumer goods which its production creates, is anti-inflationary. The smooth flow of consumer goods is dependent upon an uninterrupted supply of components and the maintenance of stocks of materials and finished goods adequate to cover fluctuations in demand. (This was the great lesson of immediate post-war recovery in Britain, and above all in what is now the EEC.) We need to encourage the growth of circulating capital, and to avoid industrial conflict which interrupts it.

A low money rate of interest, when entrepreneurs are confronted with rising wage costs and imported material prices, is essential for this purpose. And it is probably less inflationary than anything else we can do.

Economy at two-thirds cock

It is the general view of economists that low interest rates lead to an expansion of investment and thus, in inflationary times, to further inflation. This is true in a fully employed economy. In spite of low recorded unemployment the economy is operating well below its potential capacity – indeed at half, or two-thirds, cock. Costs are rising and profits are falling, with no certainty that the fall in eco-

[40]

nomic activity now commencing will do anything else but lower real incomes while prices continue to rise. Interest is a cost, and in the short run it seems to be the only cost which can be reduced without interrupting production. It will take more than a fall in the rate of interest to cause a tidal wave of investment in the present mood of the capital market. The essential thing is to relieve pressure on profits, and thus to stimulate the maximum production from existing resources.

Negative real rate of interest

It may be argued that the 'real' rate of interest is at present very close to zero, because of the declining value of money. This is of little practical interest to entrepreneurs, who must constantly increase their outlay on the replenishment of stocks if the stock of real circulating capital is to be maintained, even at its present cripplingly low level: it needs to be increased. It may be argued that it is unjust to lenders to reduce the money rate of interest to a level at which the 'real' rate is, in the short term, negative. But, if the rate of price increases can in this way be checked, the real rate of interest will commence to recover. To set the money rate at what looks like the long-term rate of return on risk-free investment constitutes only a temporary disadvantage to the lender, and will, by causing a rise in the market value of existing fixed-interest bonds, assist that sector of society which has suffered most from inflation.

In this context it is instructive to consider the traditional cure for extreme inflation, twice successfully applied in Germany; this is to demonitise the existing currency, offering one unit of the new for 'x thousand' units of the old. This appears to be horribly unjust, in the short run, to those who have suffered most from inflation because their occupations have required them to keep a high proportion of their assets in currency. But it stabilises the real value of their stock of currency, which was in process of rapid decline. Their long-term position is better than it would otherwise have been.

Rate of return on risk-bearing

What a solution requires is (a) recognition that probably a lower rate of return on the ownership of capital and (b) higher hourly earnings of most kinds of labour are the conditions which free bargaining would at present produce in normal circumstances. Also the rate of population growth has probably, for the time being, raised the real cost of food, while the growth in demand for mineral fuels combined with the recent recognition and monopolistic organisation of the bargaining power of their owners, has slowed down the rate of growth of output for the free enterprise sector. It is thus doubtful if real wages can rise much, and probably certain that the long-term rate of interest must fall. But if the free enterprise sector is to operate, the rate of return to entrepreneurship and risk-bearing must rise, as uncertainty increases.

At present, as world-wide inflation has developed, purely short-term forces dominate all bargaining, and this has tended so to over-emphasise the natural long-term trends that the rate of return on the ownership of capital is now probably negative, and profits are dangerously low. If inflation is to be checked, some factor or product prices must be held constant, or undergo a decline, during the period in which past price increases work themselves out.

International agreement on interest rates

If this is done by controlling the prices of products, profits, already dangerously eroded, will decline still further. If it is done by any form of wage freeze, strikes are likely to interrupt the supply of consumer goods and intensify the rate of inflation. (This is not to deny the enormous importance of the utmost restraint in wage claims, nor even to admit that threshold agreements guaranteeing the existing level of real wages are consistent with current reality.) But, at least in Britain, it seems probable that the price which the unions would exact for a comprehensive control of wages would be a central control over all incomes, indeed a reduction almost to nil of all incomes except wages, salaries and pensions. This would be tantamount to introducing the dreary form of bureaucratic socialism anticipated (without enthusiasm) by Schumpeter. It would not be a means of reviving free enterprise, but would be equivalent to its supersession.

In these circumstances, it appears that the one thing which can be done is to effect a drastic reduction in the money rate of interest. No one country can do this alone: what is needed is a re-introduction of something like the tri-partite agreement of 1936,[1] combined with

[1] Britain, France and the USA agreed to hold exchange parities with provision for re-adjustment.

provision for an expandable fund of liquidity, which will fix agreed rates of parity for the major currencies, with agreements to cover their adjustment should they become unrealistic, and to reduce and hold down the basic rate of interest. This would break the circle at the only possible place, and lessen the cost of borrowing from the oil producers.

Restraint of state expenditure
The level of state expenditure on anything but aiding the production of consumer goods must come down: all such expenditure creates money incomes without adding to the supply of goods upon which income is spent, and is inflationary. To counter it by increased taxation (which must fall on the consuming masses if consumption is effectively to be restrained) is liable to increase wage claims. Indeed restraint upon most forms of industrial investment (already imposed by our present uncertainties) should not be removed too rapidly, because until the new plant comes on stream this too would be inflationary.

Britons Must Consume Less!
SIR ALEC CAIRNCROSS
Master of St. Peter's College, Oxford

THE WORLD is faced with two distinct crises. The first is an employment crisis brought on by fear of inflation and reluctance to pursue
expansionary policies while costs and prices are rising fast. This
crisis is more apparent abroad, but Britain cannot escape it since her
enormous external deficit limits very severely her room for man-
oeuvre. The second is a crisis of world imbalance associated mainly,
but by no means entirely, with the problem of financing imports
from the oil-producing countries. Flows of capital are taking place
on an unprecedented scale from inexperienced lenders to countries
that were hitherto capital exporters. The resulting accumulation of
short-term debts on a vast scale in the middle of a world recession
could greatly aggravate the employment crisis. Neither crisis is of
Britain's making but Britain is highly vulnerable and peculiarly
unprepared.

Her vulnerability is both economic and political. It is economic
because of her dependence on imports of primary products, lack of
reserves, and enormous deficit. These narrow the scope for effective
action. It is narrowed still further by political factors: the lack of
authority enjoyed by government, the destruction of the credibility
it requires for the exercise of leadership, the absence of any assurance
of unity. The public is unprepared for setbacks and has expectations
of increasing affluence unrelated to past performance.

Prospective world depression

Although the world depression in prospect is unusually severe by
post-war standards, things are not likely to get out of hand. In the
course of 1975 expansionary measures are likely to be taken in the
main industrial countries which, even if not immediately effective,
should check the downswing and make possible a resumption of
growth in 1976. It is also possible, and indeed probable in the light
of experience in 1974, that a way will be found of channelling the

surplus revenues of the oil producers through bilateral and multi-lateral arrangements in such a way as to allow the international payments system to work reasonably smoothly.

What distinguishes the British situation and makes it much more alarming than the international is the persistence of a number of major maladjustments of which three are of special importance:

(i) *The size of the external deficit*

There has so far been no great difficulty in financing a current account deficit of about £4,000 million and no major foreign exchange crisis has occurred. This has lulled many skilled commentators into a complacent expectation that things will go on in 1975 and 1976 just as uneventfully. What they overlook is that there is plenty of bad news still to come. To *enter* a period of crisis with a record deficit is highly dangerous unless there are good reasons for expecting the depression to improve the UK's balance of payments. There are reasons for expecting just the opposite. Whatever the official forecasts may show, world trade in 1975 will be anything but buoyant and could contract. While the competitive position of British exporters may be improved by a falling pound, the short–run consequences of depreciation, as was apparent in 1968, may not be favourable to the current balance, particularly if demand pressure is greater in Britain than abroad. The unemployment figures suggest that other industrial countries have more unused capacity than Britain and this can hardly be very propitious for the balance of payments.

(ii) *The extinction of profit margins*

It is particularly alarming that, before demand has slackened more than marginally, profits have virtually disappeared. The nationalised industries, when last heard of, were making a collective loss of £1,000 million or more; and in the private sector British manufacturing companies in the first half of 1974 made barely enough in profits, after providing for depreciation, stock appreciation and taxes, to meet the cost of wages and salaries. Employment incomes are rising past the limits set by output, squeezing out profits and leaving a large external deficit. It goes without saying that this is not a sustainable position: whether industry is private or public it must yield a surplus to survive.

[45]

(iii) *The high level of consumption*

Thanks in part to the elimination of profit margins, consumption has been maintained and expanded during a period when import prices might have operated powerfully in the opposite direction. Between the middle of 1972 and the middle of 1974 the price of imports rose by over 25 per cent in relation to the price of exports. Since imports are roughly 20 per cent of national income, the net loss of income from this single source was over 5 per cent in two years. Yet consumption rose faster in 1973 than in any recent year and in the first half of 1974 was nearly 5 per cent higher than two years previously. The consumer, subsidised out of borrowings and taxes, shouldered no part of the burden of higher import prices; but business, caught by price controls, contributed through loss of profits.

How did this come about? The answer lies partly in the efforts of the Conservative Government to spend their way out of an earlier depression and the stimulus that this gave to consumption, and partly in acceptance of the view that external deficits, in the absence of a real transfer of resources to the oil exporters, were inevitable and desirable. But there was no need to concentrate any fiscal stimulus so exclusively on *consumption*. The increase that has taken place is equivalent to half the external deficit; and it would be no bad thing if the British deficit had kept within a limit of £2,000 million rather than £4,000 million. It is striking to see the contrast over the last two or three years between the experience of Britain and Germany. The rise in Germany's GNP was not so very much faster; but she moved into overwhelming surplus in the same years as Britain moved into overwhelming deficit. The main difference lay in the use to which the two countries put their additional resources. Germany put most of the addition into exports, Britain into consumption.

Prospects for 1975

As the year proceeds the government is likely to be faced with a deepening crisis:

(a) Unemployment will be rising fast, the external deficit will be very high and may be increasing, and the power of the Government to operate successfully on either will be in question.

[46]

(b) If prices rise faster in Britain than in 1974 – an expectation shared by few if any other industrial countries – the pound will continue to slide and this may arouse the fears of large holders of sterling. There could then be an acute exchange crisis.

(c) Any depreciation of the exchange rate will aggravate the rise in prices which, given the current trend in wage settlements and the need to restore profit margins in the public and private sectors, is bound to be rapid and unlikely to slacken for at least some months.

(d) Political events at home and abroad may intensify the crisis. Abroad there is, for example, the risk of fresh hostilities in the Near East and a consequent interruption to supplies of oil. At home there are all the uncertainties of the Common Market referendum and the instability of a government divided on critical issues on which it is committed to act in 1975.

Three questions government must face
The three overriding questions which the Government has to face are:

1. How can it assure itself of the means to cover the external deficit either by direct borrowing from the oil exporters or with support from the IMF or any other agency charged with maintaining good order in international finances? This is perhaps the easiest of the questions to resolve provided holders of sterling balances do not become alarmed by the trend of events or the drift of policy and start withdrawing their funds. About half the deficit on current account is covered already by an inflow of long-term capital from abroad which reached about £2,000 million in 1974 (largely for North Sea oil development). Public authorities are borrowing extensively in the Euro-bond market with the help of Treasury exchange guarantees. The Government has also arranged substantial lines of credit that have not yet been drawn upon.

2. How can the external deficit be reduced to a more sustainable figure, say, £2,000 million? This is in some ways the most difficult question of the three. For a time it looked as if the Government's attitude was to trust to luck and rely on a change in the terms of trade to reduce the import bill. Export prices did move up sharply; but the effect was to remove most of the competitive advantage left over from earlier devaluations. (It is conceivable but unlikely that British

exporters are merely taking higher profits on an unchanged volume of exports until they can market additional supplies.) While the terms of trade in 1975 are likely to show some improvement on the average over 1974, the main improvement in the current account must come from changes in volume. There seems no escape from the conclusion that the exchange rate will have to go on slipping so as to restore the competitive position of British producers in foreign and domestic markets. At the same time, conditions within the UK will have to be such as to facilitate a powerful swing in the trade balance.

3. This brings me to the third question: how can consumption be reduced in relation to other claims on resources? Even to cut the external deficit in half without any change in output would mean a major shift between competing claims: if consumption took the whole shock of adjustment it would have to fall by 4 per cent. So large a reduction in a single year (even measured in relation to trend) is unheard of since the war. If, simultaneously, output were to fall, which is by no means out of the question, the reduction would be even larger. No doubt consumption will not take the whole shock, but it must take most of it, particularly when there is a need to expand investment. The longer we delay the necessary adjustments the more difficult they will be.

Lower consumption, rising unemployment
Adjustment must mean higher prices and the risk of an aggravation of wage-inflation. But this is a risk we cannot avoid. It will not be eliminated by higher unemployment, as some economists seem to assume, but it will certainly be reduced in an environment of *rising* unemployment such as cannot but exist in 1975.

The Government has already taken the first step to allow some upward adjustment in prices by relaxing the more draconian elements in price control and raising prices in the nationalised industries. It needs to carry the process further while avoiding putting the full weight of higher prices on the lowest-paid workers. At the same time the Government needs to expand investment, not necessarily in the private sector but above all in energy and transport. If we must borrow abroad let us build ourselves some useful assets out of the proceeds and not perpetuate higher levels of consumption than our circumstances justify.

[48]

It is sometimes argued that the Government is powerless to cut consumption, and the experience of 1969–70 is cited as evidence that any temporary success merely brings on more rapid inflation later. Those who argue this way rarely go on to argue that tax cuts are equally powerless to stimulate consumption. If the Government were in fact so impotent there would be no escape, short of all-round rationing, from our present difficulties. But whatever one may think of rationing – and it is not without its uses, for example in dramatising the need to conserve energy – it would be foolish to regard it as the only way out. If government measures to raise prices and reduce consumption did provoke a wage explosion, the inevitable consequence would be a steeper rise in unemployment; and there is as yet little in British experience to justify the expectation of an *acceleration* in wage settlements and unemployment simultaneously.

A reduction in consumption does not automatically shift resources into exports: it may merely increase unemployment. It was precisely this dilemma that led to the Barber budgets. Given time and an appropriate rate of exchange, as Mr Roy Jenkins showed, the shift can be accomplished. But is there a case for speeding up the process of adjustment by introducing import restrictions? These could hardly be helpful to exports but would divert consumer demand away from imports to domestic sources of supply.

Import restrictions to be avoided
It is to be hoped that the Government will resist the temptation to restrict imports. Restrictions would inevitably be directed mainly against imports of manufactures and hence against other industrial exporters, especially other members of the EEC. This would, to say the least, stir up trouble in the middle of the negotiations in Brussels and set an example that could prove very unfortunate for a country as dependent as Britain on open world markets. It would do nothing directly to correct the fundamental maladjustment between consumption and other elements in output.

Nevertheless in a real exchange crisis with the rate falling out of control and no foreign credits to sustain it, the Government might have no option but to introduce imports restrictions as a means of rationing foreign exchange. It would have itself to blame if it allowed such a situation to arise.

The analysis and prescriptions given above make no reference to the Government's borrowing requirement or to the money supply. This is because it seems much better to start by looking at the dog than at the tail that is alleged to wag it: to examine the real changes that need to be made than the financial consequences of these changes. If consumption is cut while output is sustained by higher exports or higher private investment, the borrowing requirement will fall automatically. Fiscal policy is an inevitable ingredient in the management of the economy. But it is likely to make for confusion if fiscal policy takes shape in terms of some pre-determined target for the borrowing requirement rather than in terms of the competing claims on real resources.

Priorities in the Crisis

PETER M. OPPENHEIMER

Tutor in Economics,
Christ Church, Oxford

BRITAIN FACES no economic problems new to her. But the old ones have become bigger and more threatening. So far as the aggregate balance of the economy is concerned, the direction of change needed in the rest of the 1970s is the one that has been desirable since at least 1960: on the output side, more exports and investment, less private consumption; on the income side, higher business profits, and a larger share of profits in national income.

Achievement of these objectives depends on three things: world economic conditions; British government policy; and the response of the population – especially business managers and the industrial labour force – to the first two. It is important to emphasise the limits, in a liberal society, on what the Government can do. It can set the scene, create incentives and try to improve public understanding of what is required – but no more. Its contribution is asymmetric. Wrong policies can bring disaster. Right policies can usually avert disaster, but they cannot by themselves ensure blazing success. That depends on the economy at large.

World recession, world inflation
World economic conditions are not, of course, easy at present. There is recession, probably the most serious since World War II (though that is not to say anything very dramatic, especially after the record boom of 1972-73). Despite recession, wages and prices are still rising rapidly in most countries. And there are huge payments disequilibria and international financing problems associated with oil. Britain's export performance, however, has been inadequate even in good years, and her industrial investment has been relatively weak for a decade. It is the same old story now. The majority of other industrial countries – notably Japan, Germany, Benelux and Sweden – are coping much better than Britain with the balance-of-payments

effects of higher oil prices, and are also gradually reducing their rates of inflation.

To highlight only one aspect, Britain appears already to be losing her hitherto substantial share of OPEC markets. In the first half of 1974 the dollar value of French and German merchandise exports to OPEC countries was 50-60 per cent higher than in the first half of 1973. Britain's increase was 22 per cent.

The Chancellor of the Exchequer has repeatedly emphasised that the balance-of-payments deficits which are the counterpart of the oil producers' surpluses must in the aggregate be accepted voluntarily and not fended off. Attempts to fend them off – by deflation, competitive devaluation and/or trade restrictions – may redistribute them among the debtors, but will reduce them in the aggregate only by turning recession into slump and thereby cutting world demand for oil. This is true. But it is not up to the United Kingdom to be world economic policeman. Our first job is to put our own house in order. And the Government, though it cannot do the job on its own, must give the lead.

Overseas indebtedness

Two priority issues stand out. First, the UK is currently running up debts to foreign countries at an annual rate of about $5\frac{1}{2}$ per cent of GNP. This is the result of reckless financial expansion under the Conservatives in 1971-73, as well as the jump in oil prices. The borrowing is serving largely to sustain private consumption. Private industrial investment dropped by a quarter between the end of 1970 and the end of 1972, recovered by 7 per cent in 1973 and since then has been falling again.

The Government has expressed concern about the weakness of investment, but has tended to be complacent about foreign borrowing, because of the prospect of North Sea oil. This calls for two comments. First, insofar as North Sea oil is sustaining our creditworthiness and therefore our current consumption of goods and services, its benefits no longer lie in the future. They are already being taken. The net benefit in the 1980s will be only that which remains after providing for service of the debts currently being incurred.

Secondly, North Sea oil is expensive to extract. Its production costs (including interest on capital) are currently estimated at about $5 a barrel, compared with a world market price of $10. Self-sufficiency in oil would then yield Britain a real-resource saving (collected by the government in tax and royalties) of only half her present import bill, i.e. some £1,750-2,000 million per annum. This is about 2½ per cent of GNP. Account must then be taken of the debt service charges mentioned above. It is difficult to forecast how large they will be in real terms by 1980, but it can hardly be less than 1 per cent of GNP. Thus, a net gain in real GNP of 1½ per cent from self-sufficiency in oil by the early 1980s would be a good result. (This, of course, is by comparison with 1974; relative to October 1973 Britain would still be worse off, because North Sea oil is so expensive to produce.)

If the world price of oil declined or if North Sea production costs rose further, the gain from self-sufficiency would be progressively smaller. On the other hand, more would be gained if price/cost relationships moved in the opposite way, or if, at present prices, Britain became a net exporter of oil. The net gain per barrel of oil exported would be the difference between selling price and production cost. Taking these still as $10 and $5 respectively, Britain would need to produce half as much again as her own consumption in order to cut her net oil bill in terms of real resources back to the pre-October 1973 level of £800 million per annum. This would mean a further boost to GNP (on top of the net gain from self-sufficiency) of about 1½ per cent. These are all very modest figures. They cannot possibly justify borrowing abroad on the present scale for a period of years in order to sustain domestic consumer spending.[1]

Business profits and liquidity

The present weakness of industrial investment is partly a direct result of the squeeze on business profits and liquidity – the second priority issue. Profits always suffer relatively in recession, because of the heavier burden of overheads (including a lot of labour costs) at times of below-capacity working. There was a severer squeeze

[1] Reporting to the House of Commons on the EEC summit meeting in December 1974, the Prime Minister asserted that North Sea oil would make it appropriate for the UK to pay a larger share of the EEC budget in the 1980s, assuming that we remain a member. (*The Times*, 17 December, 1974.) EEC budgetary contributions are supposed to be loosely linked to members' GNP. If North Sea oil raised the UK GNP by 3 per cent (which, as noted above, would be a considerable achievement), this would put up the UK's share of EEC GNP in the early 1980s from something like 17 to 17½ per cent, a trivial difference. Is the Government living in the realm of fantasy with regard to North Sea oil?

than usual in 1969-71, because of the combination of recession and rapid wage-price spiral. This time the inflation is even worse; company earnings have been held back by price controls since the autumn of 1972; and the economic climate is permeated by exceptional uncertainty which was not present in earlier periods.

The impact of these factors has been compounded by the established systems of accounting and taxation, which take no account of inflation and which treat stock appreciation as equivalent to trading profits. The seriousness of the position is indicated by the fact that

'Net of stock appreciation, profits as a proportion of current price GDP fell to only $5\frac{1}{2}$ per cent in the first half of 1974, compared with nearly 15 per cent in 1963. At the same time the proportion of stock appreciation in gross profits rose from $3\frac{1}{2}$ per cent in 1963 to more than 60 per cent . . . '[1]

There has been some debate about the status of stock appreciation. A. J. Merrett and Allen Sykes have maintained that it in no sense contributes to profitability and should be completely excluded from profit totals.[2] This is untenable. For stock appreciation adds to a company's net worth; and, as the National Institute puts it,

'From the position of tax equity, an increase in the net worth of one company constitutes an increase in its ability to pay corporate taxes relative to another company which had not experienced a similar rise in asset prices.'[3]

However, the conventional view which Merrett and Sykes were combating and which was subsequently defended by Wynne Godley and Adrian Wood is also unacceptable. The reason is implicit in Godley and Wood's description of the conventional approach:

'The procedure adopted by accountants . . . is to add to the expenditures made in the period the opening value of stocks, which in effect measures those costs incurred in previous periods with respect to goods sold in the period in question. At the same time, they subtract the closing value of stocks, because this in effect

[1] *National Institute Economic Review*, November 1974, p.10.

[2] 'The real crisis now facing Britain's industry', *Financial Times*, 30 September, 1974.

[3] *National Institute, loc. cit.*, p.22.

measures those costs incurred in the current period with respect to goods which will be sold in future periods . . .'[1]

In other words, a sale to oneself-in-the-next-period is treated as equivalent to an arms-length sale in the current period. This assumes that there is no uncertainty about the future ('goods which *will* be sold . . . '). In the presence of uncertainty the equivalence disappears. At normal times, uncertainty is not too great and the amount of stock appreciation is a small fraction of trading profits, so the conventional assumption is unobjectionable. At times like the present this is surely not the case, and something less than the whole of stock appreciation should be counted as profits.

Inflation accounting and uncertainty

In principle, this problem of uncertainty is separate from that of 'inflation accounting'. In practice, the two seem bound in large measure to overlap. If the general price level were stable, or if a full-scale system of inflation accounting were in force, with replacement-cost depreciation, indexed tax rates and so forth, it is scarcely conceivable that firms would experience a large taxable appreciation of stocks combined with exceptional market uncertainty such as exists at present.[2]

Be that as it may, the present extraordinary ratio of stock appreciation to trading profits indicates, in part, a genuine profits problem and not merely a liquidity or financing problem. This helps to explain the hair-raising collapse of the UK equity market. Investors evidently do not believe that the government is prepared to let inflation be adequately reflected in business profits in the foreseeable future. The maintenance of companies' real capital will (on this view) be called in question, and dividends become an increasingly unreliable form of income.

[1] Godley and Wood, 'Uses and Abuses of Stock Appreciation', *The Times*, 12 November, 1974.

[2] There is even a conceptual overlap. Inflation accounting is usually concerned to enable companies to maintain the *real* value of their capital intact before becoming liable to corporation tax. To an economist, however, the concept of real capital is essentially forward-looking: it is the discounted value of a future stream of earnings, rather than the hypothetical present cost of a machine bought at some time in the past.

I am grateful to my colleague Michael Bacharach for discussions on the question of uncertainty and profits.

Restrain home consumption: remove control on profits and prices
What, then, should be done? The Government must maintain a restrictive stance towards home demand in the aggregate. But the present mode of restraint puts too much pressure on company finance and not enough on private consumption. This tends to maximise the fall in investment and probably also the amount of personal hardship caused. Eventually, bankruptcies will make, say, 10 per cent of the work-force unemployed; and their consumption will be cut by 25 per cent, giving an aggregate cut in consumption of $2\frac{1}{2}$ per cent. It would be fairer, and less risky to business, to try to enforce a cut of $2\frac{1}{2}$ per cent (or a bit more) in everybody's consumption. This calls for the removal of substantially all control on profits and prices, and for some increase in personal taxation, preferably VAT.

Raise taxes and unemployment
It is arguable that there should be a further rise in taxation to finance increases in public expenditure. At present, parts of the public sector, especially the social services, are being forced to bear a disproportionate share of cut-backs. Some re-allocation of spending within the public sector should also be possible – cuts in local authority staffs, for instance, and in road maintenance (our urban and secondary roads are over-maintained anyway).

Some rise in unemployment and in short-time working is inescapable. But it need not be substantial if British industry, with support from the unions, steps up its export efforts – to the OPEC countries among others.

The Government has begun to move in the general direction suggested above. The November 1974 budget relaxed the price code and postponed corporation tax on the bulk of stock appreciation for the year 1973-74, with similar or greater relief promised for 1974-75. It also announced the gradual withdrawal of subsidies for the nationalised industries. Just before Christmas commercial rents were de-controlled, hurriedly and somewhat offhandedly, as from February.[1] And the price of school meals was raised.

Tougher approach required
This raises the question of presentation and speechifying. Some of

[1] The announcement was made in the form of a Written Answer by the Secretary of State for the Environment just before the parliamentary Christmas recess (Hansard, 20 December, 1974, col. 609).

the measures already taken, and more of those that will need to be taken in the period ahead, are unpopular. They need to be announced in a tougher way, supported by more explicit argument. In particular, it needs to be emphasised, first, that 'protection of the weak' from cuts in spending power does not mean protection of the average wage-earner; secondly, that the bulk of business profits is not the income of rich individuals but the means of financing investment and therefore jobs and higher earnings for everyone in the future; thirdly, that the standard of living consists of more than the number of pints which can be bought with one's weekly take-home pay, and includes such things as the quality of hospital care and of school education; and fourthly, that the people of most other industrial nations have so far made a very much better job of coping with higher oil prices than we have, and that things will become far more unpleasant – with rationing and other restrictions characteristic of a war economy – if we exhaust our international credit-worthiness.[1]

[1] In the early weeks of 1975 Government Ministers, including the Prime Minister and the Chancellor of the Exchequer, have started to say some of these things – and quite forcefully. This is very welcome.

Last Straw or Turning Point?

WILFRED BECKERMAN

Professor of Political Economy,
University of London

THE CRISIS has been coming for a long time, particularly for the British economy. The recent developments in the oil market have merely accelerated events. The biggest mistake now would be to assume that our troubles will be almost over if we hang on for a year or two until North Sea oil begins to flow.

In the world as a whole, many developments have sown the seeds of international tensions and conflicts. These include the understandable nationalistic pressures from the newly established less-developed countries, most of which had been either ex-colonies of some great power or economically dominated by the richer countries. At the same time, the economic inter-dependence of the countries of the world has been increasing with the rising importance of international trade.

Internally, also, the changes in the labour market since the war, together with the generally increasing role of public expenditures, have set the stage on which different groups – and sometimes the same groups or individuals acting in different capacities – have been able successfully to pursue conflicting objectives. These developments have been related to the gradual acceleration of inflationary pressures in all the advanced countries over the last 20 years. Even countries such as Switzerland or Belgium or New Zealand, which had maintained remarkable records of price stability during the 1950s, have recently experienced inflation rates that are not merely high by their own standards but also by the experience of other industrialised countries.

Economic boom and rising import prices
The oil crisis came at the worst possible time for inflation. During the period 1972-73 the economic boom in the industrialised countries as a whole had got out of hand. This led to excessive pressures of

demand in the industrialised countries and to rising prices of imported primary products. Without any oil crisis a downturn in activity – automatically or induced by government stabilisation policies – would have followed, accompanied by continuing rapid inflation for a time whilst the earlier increases in domestic and imported costs worked their way through the system. Thus 1973 was already a year of accelerating inflation and upheavals in international monetary arrangements, and even with no oil crisis 1974 and 1975 would have seen continued inflation accompanied by slackening output and rising unemployment.

The oil embargo of October 1973 and the subsequent quadrupling of the oil price refuelled the inflation of primary product prices (which had been on the point of petering out); it vastly complicated the international monetary scene; and it imposed a further cut in real incomes at a time when some such cut was coming anyway, as a result of the preceding excessive boom. But some of the unavoidable effects of the rise in the oil price – as distinct from further aggravating measures that governments may take in the attempt to pursue their economic objectives – can be easily exaggerated. The impact on inflation is relatively small. The rise in the price of imported oil adds about 2.0 per cent to the GNP price level for the UK, as compared with domestically generated inflation which, for the year 1974 as a whole, was about 12 per cent over 1973 ('GNP deflator') and which is currently running at a much faster rate.

Higher investment in energy-saving industries
The automatic cut in demand is less than 2 per cent on account of various leakages, but this is, of course, rather big compared with the size of demand changes to which we are accustomed. The unavoidable technical impact on the longer-run growth of the British economy is probably very small indeed, and is not even necessarily adverse: until we have to pay for our higher oil import bill with higher exports, the increased balance-of-payments deficit could be offset by higher investment. In terms of output (as distinct from real incomes which have been cut by the deterioration in the terms of trade), we could carry out more investment which would be matched by the higher capital inflow originating in OPEC savings. There will also be some switch towards more investment in energy production and energy-saving industries, while presumably the

[59]

profitability of investment in some energy-intensive industries will be reduced. But the direct total impact on investment and the longer-run growth rate is likely to be negligible compared with the possible effects of measures by governments to keep down the pressure of demand. These will depress investment incentives in general.

Governments may do this for various reasons, including the international monetary repercussions of the oil crisis on their foreign balances, which constitute one of the most serious effects of the oil crisis. If the OPEC countries could absorb enough imports for us to be able to pay for the extra oil bill this way, our economies would have to accept the 'real' burden of the deterioration in the terms of trade that the rise in the oil price represents. But this would probably be much easier now than finding international monetary arrangements to handle the vast funds the OPEC countries will accumulate because they cannot spend their oil revenues.

For the world as a whole, macro-economic equilibrium requires that the increased world propensity to save (caused by the redistribution of income in favour of the low-spending OPEC countries) needs to be offset by higher investment or lower savings by the industrialised countries. Ideally, higher domestic investment in the industrialised countries would kill two birds with one stone. First, it would fill the gap in domestic demand caused by the higher import bill. Secondly, it would provide the basis for faster real growth in our economies so that the *real* burden of paying for the oil with exports in a few years time (when the OPEC countries will have stepped up their importing capacity) will not weigh too heavily on our resources (not to mention paying off our debts if ever necessary). Unfortunately, the whole conjunctural situation, i.e. inflation and balance-of-payments difficulties, makes this sort of obvious solution very unattractive to governments in general and to the British Government in particular. This brings us back to the longer-run state of the British economy.

Inflation and stagnation
It has now reached a position of unprecedented inflation, balance-of-payments deficit, stagnation of output and poor prospects of much investment and real growth in the foreseeable future. Nearly all the industrialised countries are experiencing similar problems to some extent, but in terms of the magnitudes involved – whether the

balance-of-payments deficit or the rate of inflation or the underlying growth rate, and so on – Britain probably has the worst combination of circumstances of any advanced country.

For many years the causes of the slow growth of the British
economy and its continued inflation have been analysed extensively.
Explanations can be of two kinds. One hypothesis is that the factors
that determine a country's rate of economic growth are numerous
and complex and not all closely inter-related. If so, they will be
randomly distributed between countries, with the result that a few
countries will be exceptionally fortunate and have a combination of
factors that leads to relatively fast growth; others will be unfortunate
and have everything against them; and the bulk will be spread over
the middle ground, some growing faster than others but with no
sharp division between fast and slow growers. The alternative hypothesis is that growth is determined by some simple model containing
one or two key variables; once one has found the key to this model
one can explain not only why some countries will grow fast and
others slowly but also specify what policies are needed to move from
the latter group to the former.

The choice between the two types of model depends, of course,
to some extent on the time-period. For example, whereas I think it
was possible to explain the observed differences in growth rates
among industrialised countries during the period of more or less fixed
exchange rates, from the early 1950s to about the mid-1960s, in
terms of foreign competitiveness, I do not think this can be an explanation of longer-term trends or even of medium-run trends in a
period when governments are no longer prepared to sacrifice internal
demand in the interests of preserving a fixed exchange rate.

For the longer run, therefore, my hunch is that the former type
of model is more likely to be correct. More precisely, I think that
the determinants of longer-run economic growth include complex
social and political influences, many of which are deeply rooted in
the historical experience of the various countries and which are not
amenable to quantitative analysis in general or economic analysis in
particular. In taking this view I am not saying anything new; I am
merely echoing the view of most eminent scholars and students of
economic growth, such as Professors Simon Kuznets or Sir Arthur
Lewis, and many others. It is a view which economists are nevertheless rather reluctant to accept, since we are trained to think in terms

[61]

of models we can handle with the tools of analysis at our disposal, and this somewhat eclectic, stochastic[1] and sociological attitude to the growth process will not fit neatly into the usual sort of economist's model. Furthermore, by the very nature of things, it is impossible to identify or measure precisely which 'sociological' (to use a short-hand approximation) factors are important and hence to test their role in the growth process scientifically. The most one can do is to show, as many investigators have shown, that it is impossible to explain growth rates in terms of quantifiable economic factors (except statistical identities, in which growth is allocated among various inputs of factors of production).

Conclusions on long-run British growth

I have digressed because this topic has a bearing on the attitude one should take to the apparent inability of the British economy to grow as fast as most other advanced countries. This alleged deficiency of the British economy has been the subject of much heart-searching; it has led to major questions being raised about the whole future of British society and the extent to which it needs to be completely overhauled in order to catch up with the growth rates of other countries. But if, as I believe, the longer-run growth rate of Britain is largely determined by sociological factors which we are, nevertheless, unable to identify, we must draw the following conclusions for policy.

First, there may be little one can do, in economic policy, to accelerate the growth rate, even assuming that much faster growth was desirable on strictly economic grounds, which is far from certain. Secondly, we cannot honestly pretend to know what social changes will lead to faster growth. Thirdly, even if we had some reasonably firm ideas about the social changes needed for faster growth, we might well conclude that the game is not worth the candle, and that we are better off pottering along in the same old sleepy way instead of emulating the vicious 'rat-race' atmosphere of some of our competitor countries. If so, the less neurotic we become over our inability to keep up with the Duponts and the Schmidts the better.

The trouble is that if we are not careful we might very soon be getting the worst of both worlds, i.e. slow growth without the cosy

[1] [Stochastic: allowing for probabilities or a range of outcomes because of uncertainty. - ED.]

[62]

tolerance characteristic of British society and that, perhaps, partly explains our slow growth. This would be so if we fail to deal with inflation, which is probably uppermost in most people's minds in talk of the current crisis. For if there is one thing likely to create social tensions it is inflation, with its attendant fears that one's standard of living might be drastically reduced if one falls behind in the race to keep up with rising prices. These conflicts would be bad enough if they reflected a struggle to maintain shares in an expanding national cake. They will be far worse if they take place – as is likely in Britain – in virtually stagnating output over the near future.

Causes of inflation – a semantic dispute?
Hence the growth problem is closely linked to the central feature of the current crisis: the rate of inflation. Unfortunately, like the growth problem, there is probably no simple solution to the inflation problem either. There can be little doubt that a high pressure of demand can add to inflationary pressures for very long periods. It is enough to recall the inflationary experience of numerous countries, where either trade unions are weak or are under the control of the authorities but where inflation has nevertheless been rampant, to conclude that union pressure is not the only factor at work. Indeed, many of the advocates of incomes policies have argued that the object of such policies is precisely to permit the economy to be run at a higher pressure of demand without having excessive inflation.

On the other hand, even the pure monetarists do not really seem to believe that the machinery of wage determination and the structure of unionisation are irrelevant. For they concede that, if the money supply were regulated in the way they would like, each country would still have a degree of what they call 'structural' or 'natural' unemployment that would correspond to the degree of monopoly in the labour market, i.e. the degree of unionisation. But, of course, if governments are committed to eliminate heavy unemployment and if this level of 'structural' unemployment is high, this merely means that its elimination in the interests of reasonably full employment will inevitably be accompanied by inflation. To argue about whether this constitutes demand inflation or cost-push inflation based on union power is then very much a semantic dispute.

For either the wage negotiation machinery must be changed to

[63]

reduce 'structural' unemployment, or demand pressures must be reduced to prevent the inflation that would accompany full employment with the existing machinery of wage negotiation. If neither is done we shall have inflation, and probably *accelerating* inflation. The acceleration of inflation over the last few years is just as consistent with the cost-push view of the inflationary process as with the demand-pull view, for expectations of rising prices can be built into any wage-push model as easily as into any demand-pull model, with the same statistical results.

Once it is accepted that the wage negotiation machinery (as well as expectations) is an important element in the relationship between employment and price stability, it is impossible to deny that here, too, we are faced with 'sociological' (to persevere with the shorthand) phenomena. In particular, Britain has a trade union structure which could hardly be more conducive to wage-push inflation, since it provides almost no effective means by which individual negotiators can be faced with the effects of their common actions on the price level in general or the fortunes of particular firms. Hence, even if trade union leaders were aware – as no doubt many of them are – that the wage negotiation system is inherently inflationary, there is little they can do about it. The leaders of each union have been elected to protect the interests of their own members, and it would be foolish to expect them to restrain their own pay claims unless they could be really certain that all other unions would show the same restraint. Otherwise they would merely see a fall in the real incomes of their members. Nor are national unions often faced with the prospect that their wage claims would put particular firms out of business, for all firms employing members of the same union will be affected in the same way, which means not much affected at all (except insofar as they experience increased competition from imports).

Shop-floor bargaining or centralised wage negotiations?
Thus, there would seem to be two possibilities. We have to move further either in the direction of shop-floor bargaining, as many authorities have suggested, or in the opposite direction towards centralised negotiations between the trade unions taken together and the employers, along Scandinavian lines. In the former, a firm faced with an excessive wage claim will be threatened with bankruptcy,

[64]

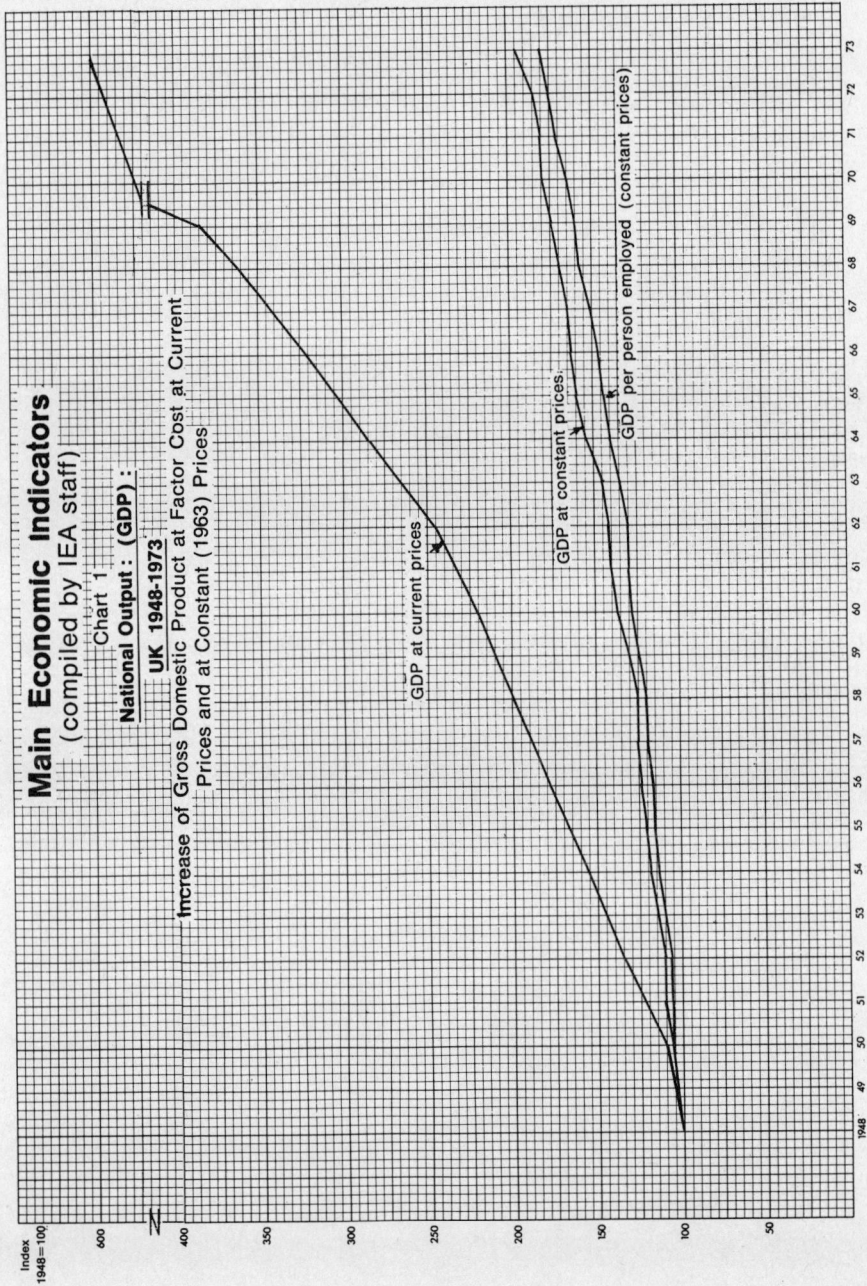

Main Economic Indicators
(compiled by IEA staff)

Chart 1

National Output : (GDP) :

UK 1948-1973

Increase of Gross Domestic Product at Factor Cost at Current Prices and at Constant (1963) Prices

Index 1948=100

GDP at current prices

GDP at constant prices

GDP per person employed (constant prices)

(i)

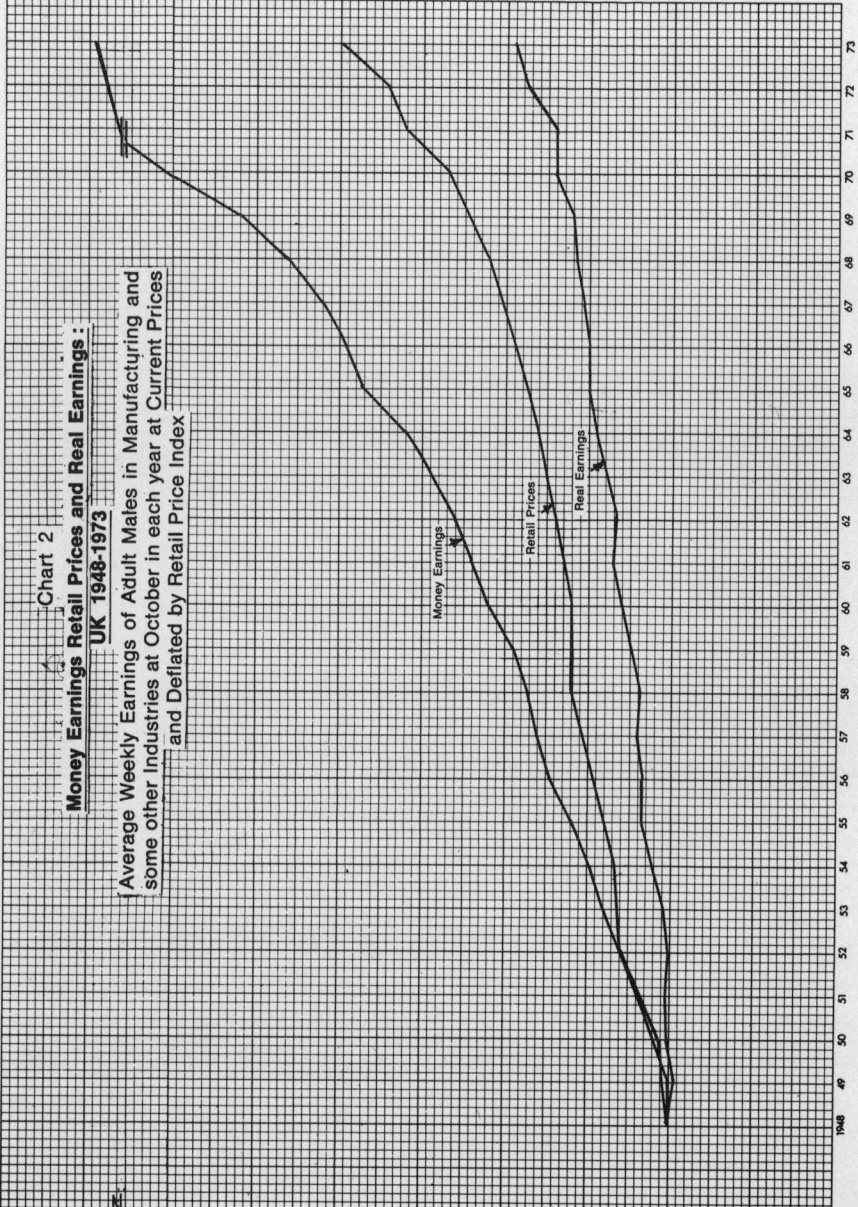

Chart 2
Money Earnings Retail Prices and Real Earnings:
UK 1948-1973

Average Weekly Earnings of Adult Males in Manufacturing and some other Industries at October in each year at Current Prices and Deflated by Retail Price Index

Index
1948=100

(ii)

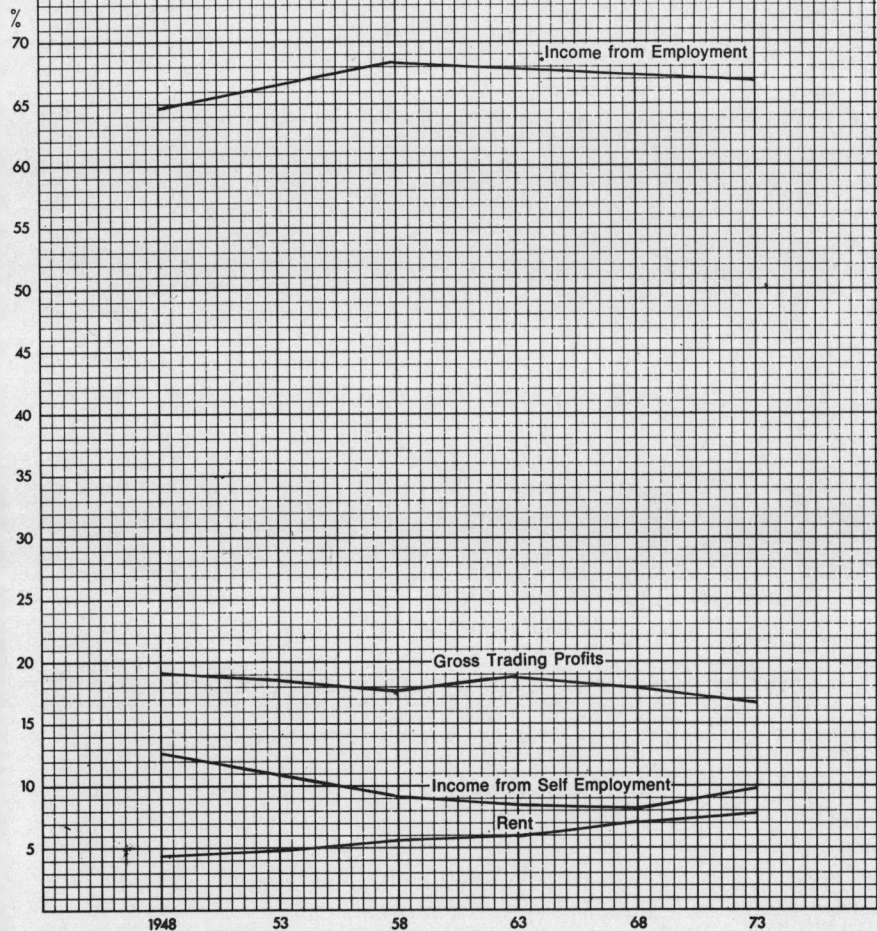

Chart 3

Shares in Total Domestic Income :

UK 1948-1973

Percent of Total Domestic Income
before providing for Depreciation
and Stock Appreciation

Income from Employment

Gross Trading Profits

Income from Self Employment

Rent

(iii)

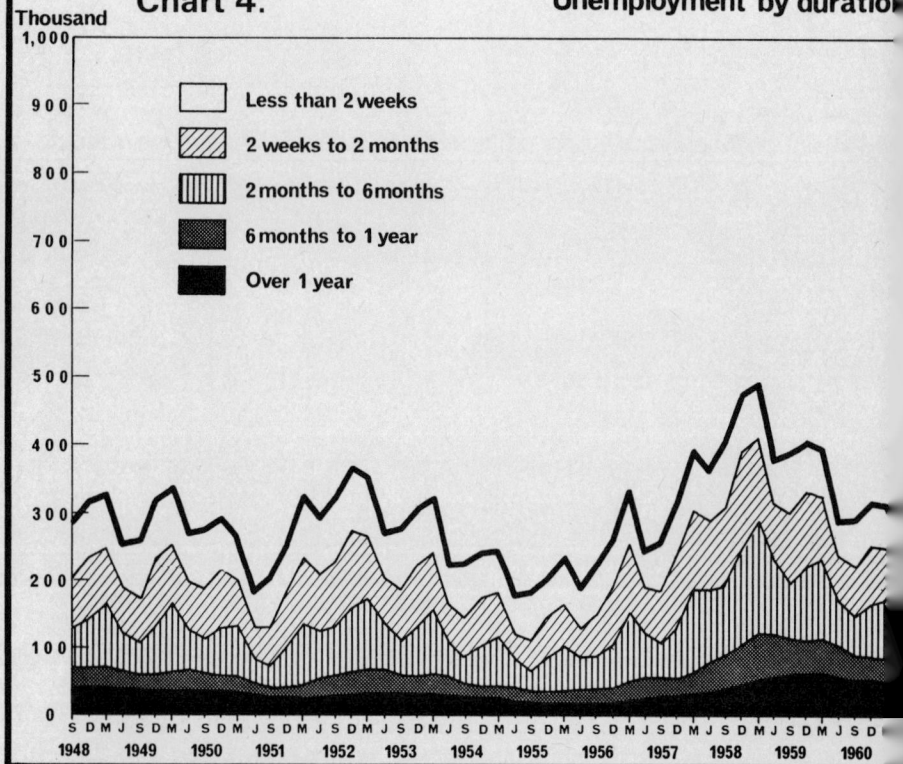

Chart 4.   Unemployment by duration

Thousand

Legend:
- Less than 2 weeks
- 2 weeks to 2 months
- 2 months to 6 months
- 6 months to 1 year
- Over 1 year

Source: British Labour Statistics,
           Historical Abstract and Year Book, HMSO, 1969,
           Department of Employment Gazette.

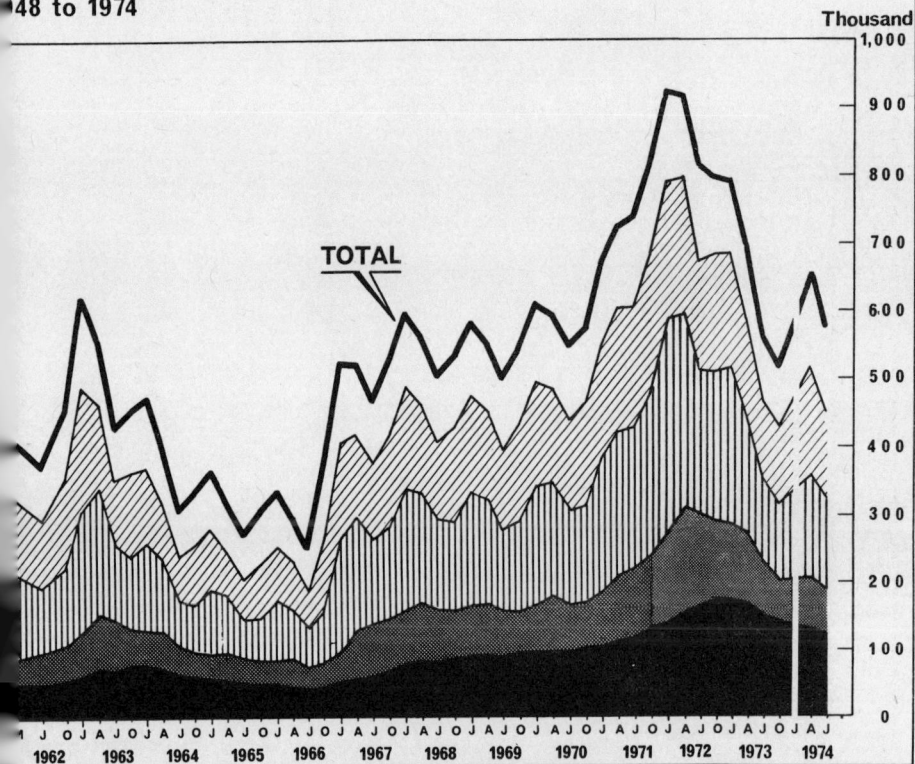

Note: The break in the Chart in January 1974 indicates the non-publication of statistics in the month of the three-day week.

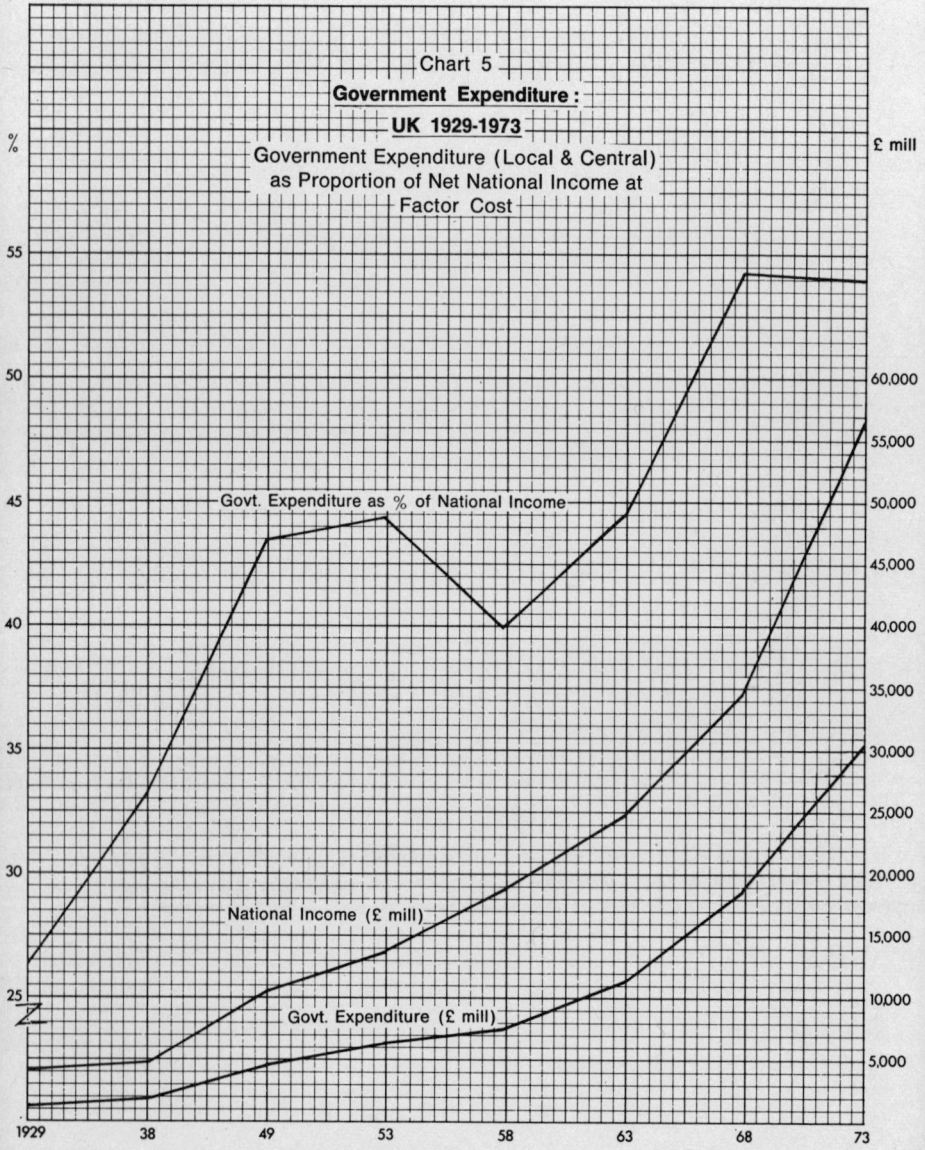

Chart 5

Government Expenditure:

UK 1929-1973

Government Expenditure (Local & Central)
as Proportion of Net National Income at
Factor Cost

Govt. Expenditure as % of National Income

National Income (£ mill)

Govt. Expenditure (£ mill)

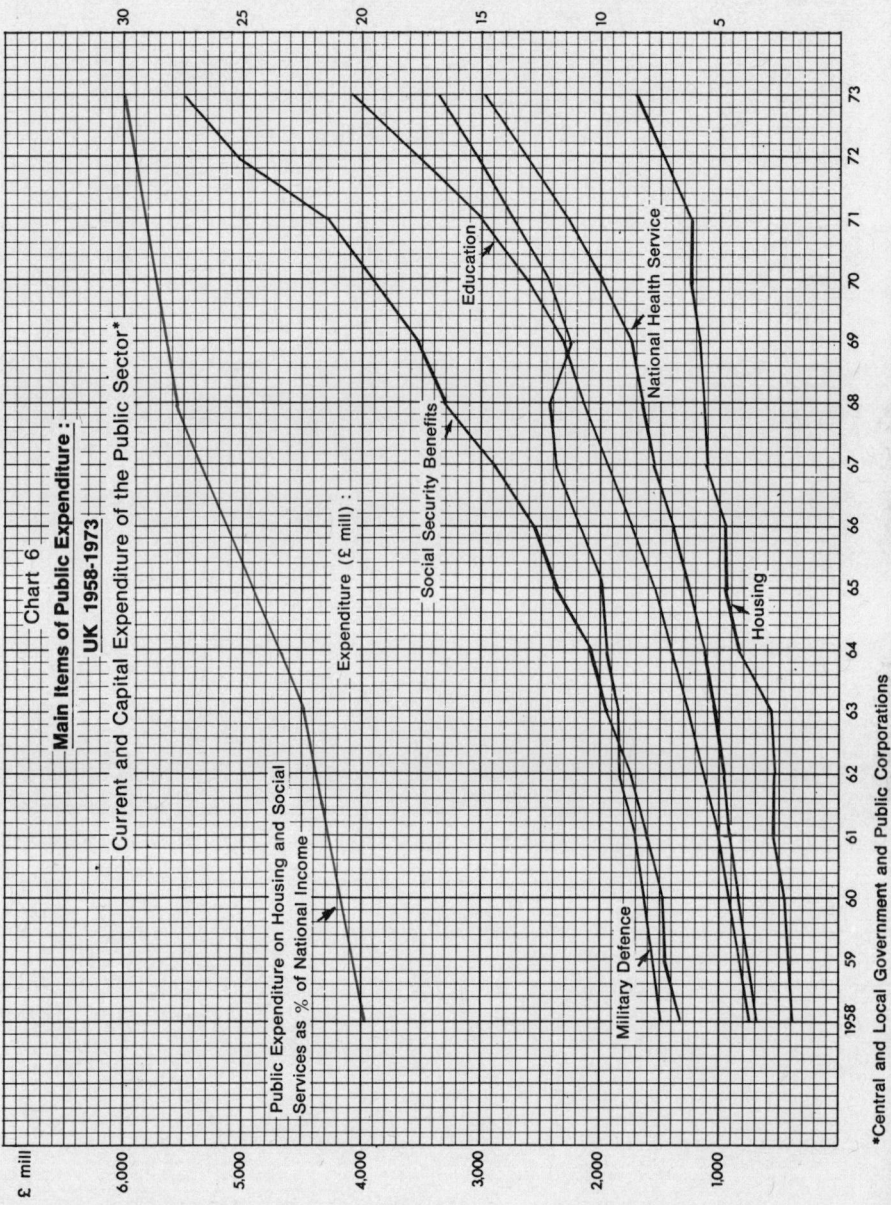

Chart 6

Main Items of Public Expenditure : UK 1958-1973

Current and Capital Expenditure of the Public Sector*

Expenditure (£ mill) :

Public Expenditure on Housing and Social Services as % of National Income

Social Security Benefits

Education

Military Defence

National Health Service

Housing

*Central and Local Government and Public Corporations

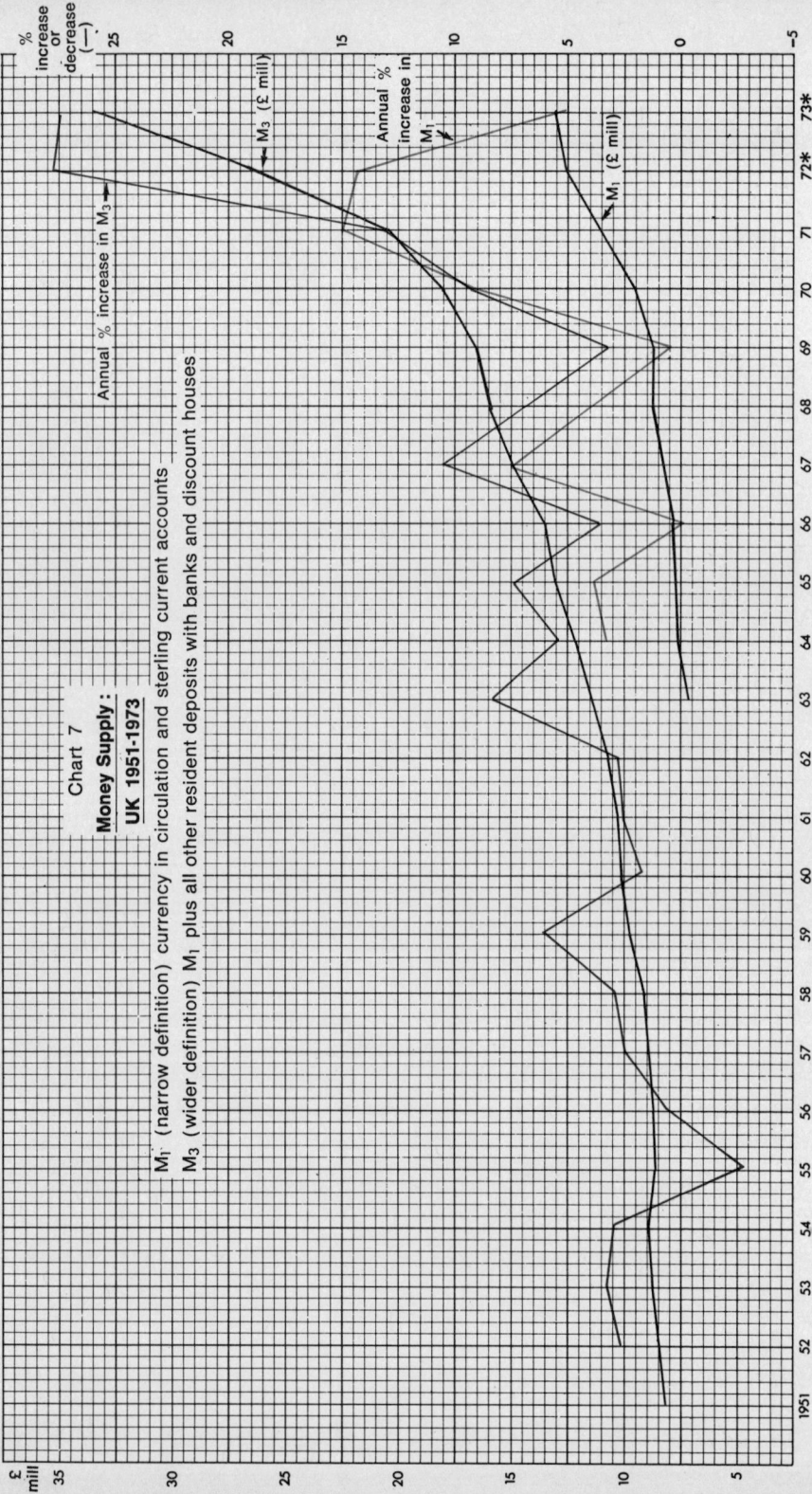

Chart 7

Money Supply :
UK 1951-1973

M_1 (narrow definition) currency in circulation and sterling current accounts
M_3 (wider definition) M_1 plus all other resident deposits with banks and discount houses

Annual % increase in M_3

M_3 (£ mill)

Annual % increase in M_1

M_1 (£ mill)

% increase or decrease (—)

£ mill

*Note : in 1972 and 1973 increase of M_3 was inflated (and M_1 increase depressed) by switching of funds into bank deposit accounts and the employees will know this since they will not be simultaneously engaged in similar negotiations with every other corresponding firm in the country. In this way they are directly faced with the consequences of their actions. In the latter – centralised negotiations – if the unions as a whole are asking for, say, 20 per cent increase in wages, they know that this implies nearly 20 per cent increase in prices so that, again, they are directly faced with the futility of their collective action.

Whatever the precise procedures the principle to be applied is that machineries have to be devised to minimise the extent to which individuals or groups can successfully pursue objectives that are mutually incompatible and that will create conflicts resolved only through rising prices.

The manner in which individual trade unions are caught in a trap which ensures that they can pursue successfully wage claims that merely lead to rising prices and to little rise in real incomes is merely one example of this basic feature of society's decision-making machinery. There are other examples, such as the way that, in their capacity as voters or recipients of health or education, the public elects governments committed to certain public expenditure programmes – and hence to the taxes to finance them. But in their capacity as individual sellers of their labour or their goods in the market most of them manage to pass on the resulting increase in taxes in higher prices or wages or profits in the generally successful effort to avoid the burden of the public expenditures they have supported as voters.

Government reform of wage negotiating machinery

In other words, the inflationary process is one example of the way in which (i) different groups in society can successfully pursue conflicting policies, or (ii) the same individuals can pursue inconsistent policies when operating in different capacities, e.g. voters or taxpayers, household bread-winners or trade unionists, etc. It is only the government that can try to reform the machinery of wage negotiation in such a way as to moderate the particularly acute conflicts that result from the archaic and clumsy British wage negotiation machinery.

It is absurd to blame trade unionists, however militant they may be – even those whose ideological motivations may be such that they are quite happy to see the whole economic system in this country collapse. For the trade union leaders have not stood for election on any promise to look after the interests of the nation. They claim, usually quite rightly, to be relatively little men who seek only to protect the interests of their own members. Hence, they have little moral responsibility for the problems of the economy as a whole. It is the politicians, who voluntarily stand for election and put themselves forward as being able to solve national problems, who have the primary responsibility, followed by our captains of industry who are paid to handle the problems of their industries, including the problems of labour relations, and who, with a few exceptions, have hardly had a new idea on the subject for the last 50 years.

The turning point?

Thus, the present crisis could be acute enough to make politicians really face up to their responsibilities with courage and imagination; in which event, instead of being the last straw to break the back of the British economy, it could prove to be the turning point. Instead of destroying some of the most valuable features of British society in a vain attempt to emulate German or Japanese growth rates, we might take advantage of the current crisis atmosphere to build on these features of society so as to help overcome more deeply-rooted problems, particularly inflation. Insofar as a major change in the wage negotiation machinery is needed, a perfect system is of course unattainable, and even a very good approximation will not be produced rapidly.

Government action in 1975

Meanwhile, the government should take the following steps.

1. It must stop rescuing bankrupt firms and must allow the pressure of demand to remain low – and probably fall even lower – for a time.

2. It must make it perfectly clear that its long-run strategy is to keep the pressure of demand low until a satisfactory wage-negotiation system is operating.

3. It must rapidly legislate some drastic provisional changes in the system, in spite of inevitable heavy opposition from many of its own supporters. There is no time to repeat the whole process of

consultation and negotiation it attempted in 1968 and 1969. It must make it clear, however, that it will start serious discussions about ironing-out the inevitable deficiencies and inequities that a rapidly devised system will entail and that improvements will be implemented as soon as inflation has been brought under control and subject to the discussions leading to an effective system.

4. It should also introduce some quick make-shift indexation of wages, pensions, taxes and savings. Again, it would take years and two Royal Commissions to produce a lovely sophisticated system but only a week to produce an adequate workable system that would suffice for the time being.

Wishful-thinking on high

But it does not look as if the Government is yet in a mood to do any of these things. It seems to prefer to go on hoping that something will turn up and that we shall muddle through. It has a faith in the Social Contract or some such miracle that resembles Chamberlain's faith in Hitler's word. History will probably judge each piece of wishful-thinking as equally incredible. Which will turn out to be the biggest disaster is still not clear.

Realities Behind Monetarism
PAUL BAREAU
'University of Fleet Street'

WITH THE social contract in tatters and inflation galloping at an unprecedented rate, the blame for the debauch of the currency is still being widely put on an excessive expansion in the money supply. Inflation is of course accompanied by an undue increase in currency and credit. But to propound this proposition is not the same as saying that we have here the whole explanation and, therefore, the whole solution of the problem. To state the proposition in these terms is to ignore the true substance of the case, the motives and objectives that lie behind the increase in the money supply. To focus on the proximate cause is to take our eyes off the main culprit, namely, the cost-push effect of increases in money earnings that are now ranging vastly beyond the expansion of productivity. Lord Robbins in the *Financial Times* said recently:

'It is not to be thought of that in a civilised and non-totalitarian society the value of money is to be left for ever to the whims and fancies of monopolistic associations, asking for any increased rate of income which comes into their heads and then, in a *milieu* of elastic credit, forcing the dire consequences of inflation on all who are not in their fortunate position.'

It ought not to be thought of; it should be inconceivable; but, alas, it is happening, and at a pace which has recently been accelerating.

Money supply a 'passive factor'

The rate of increase in the money supply, although it is a self-evident companion of inflation and may well appear to be its *proximate* cause, is in practice a passive factor. It receives its potency from *other* forces. It adjusts itself to influences bearing upon it from *outside*. Three can be most clearly identified. The first is the conscious attempts by succeeding governments to maximise the rate of economic growth and to adjust monetary and fiscal policies to this end. The second is the exercise of power by organised wage- and salary-earners to demand and secure increases in money income that considerably outstrip the concurrent increases in productivity and output. The third is the pathological fear, which has gripped all recent UK governments, of a rise in unemployment, even the short-term, reasonably remunerated unemployment which must on occasion occur in any dynamic, evolving and healthy economy.

The fallacy underlying the widespread belief that economic growth and an expansionist monetary policy go together is that the real sinews of growth are provided by intelligent and efficient management, market-oriented investment in new plant and equipment, readiness on the part of workers to operate the new plant, good human relations in industry, mobility of labour, absence of restrictive practices through cartels of employers or of demarcation rules and the whole paraphernalia of Luddite practices of employees. These are the real factors making for growth. The danger lies in assuming that the appropriate monetary policy can be a substitute for them in the expansion and prosperity of the economy.

In March 1972 Mr Anthony Barber, then Conservative Chancellor of the Exchequer, told us in his Budget speech that the rate of expansion in the money supply would be sufficient to ensure the achievement of his objective of a 5 per cent growth in the economy. If the real and basic factors in the growth equation are not there – as they certainly were not in 1972 any more than to-day – the pursuit by monetary and fiscal measures of a rate of growth which is unattainable, will merely create the short-lived consumer boom, the mounting balance-of-payments deficit and the kind of steadily worsening inflationary mess in which we have since found ourselves.

The seeds of hyper-inflation
That was when the seeds of the prospective hyper-inflation were sown. The money supply increased at a frightening rate, the borrowing requirement for 1972-73 rose to over £4,000 million, but the Chancellor and his henchmen at the Treasury devoted speech after speech to the proposition that a tighter grip on money supply would choke the economy. It was in due course to be choked by the much more direct effect of industrial anarchy, and of the political and social frictions of which inflation is the main begetter.

The most powerful of all the forces operating directly on the

[69]

value of money is the rise in money incomes. Incomes from employment represent more than 70 per cent of the Gross National Product. Changes in the level of these incomes are largely determined by the monopoly power of the trade unions. Despite the Social Contract, these powers are wielded by techniques of militancy which are relatively new to this country: disregard and open contempt of the law, readiness to blackmail the community by withdrawing essential services and dissuading or preventing others, not direct parties to a dispute, from doing their work. The legal dispensation to exercise these powers will be monstrously enhanced under the present Government's Trade Union and Labour Relations legislation.

Faced with excessive wage claims, backed by strike action, employers have been inclined to give way. The retreat has usually been led by nationalised industry and the government sector, where the inducements of success and the sanctions of commercial failure do not apply. The private sector has been scarcely less permissive, for a variety of reasons. In capital-intensive industries the cost of a strike would far outweigh the additional wages bill. There is the fear of direct militant action by the work-force, leading ultimately to liquidity difficulties, the dependence on and intrusion of government money, and the threat of 'backdoor' nationalisation. The wage restraint aspect of the Social Contract is thus being flouted on both sides of the negotiating table. Employers have been encouraged to appease unions by the promise of looser operation of price controls.

'Exchange permissiveness'
Where export industries are concerned, the necessary reassurance and mood of appeasement by employers have provided the doctrine of exchange permissiveness which now holds sway. In the Budget speech of March 1972, we were told that 'never again would the growth of the economy be held back by attempts to defend an unrealistic rate of exchange'. The first application of that dictum was to be provided by the decision to float sterling taken three months after the speech was delivered. The attitude persists and underlies the belief that, when in balance-of-payments trouble, sterling will be allowed to depreciate. In consequence export industries, which should be those most subject to the discipline of competitive costs and prices, have also been inclined to give way to exorbitant wage

[70]

demands in the hope that a depreciation of sterling against other currencies would 'save their bacon'.

In an inflation of labour costs, unemployment should increase and has done so to a small degree. In many enterprises the labour force has been slimmed as a direct result of the rise in labour costs. The increase in unemployment should have provided its own automatic check on wage inflation. The operation of basic natural laws has been hampered by the intrusion of other forces. The 'Paish principle' – the belief that with unemployment at $2\frac{1}{2}$ per cent, inflation would be halted in its tracks – has gone right out of the window. What has destroyed it is not only the monopolistic power of the unions but the speed and generosity with which social security services plus redundancy pay, plus unemployment benefit, have come to the rescue. Britain is the only country in the world where the government finances strikes.

Banks have contributed to wage inflation

How do wage-induced and cost-push inflationary forces affect the money supply? To serve the cause of growth and full employment, monetary policy has over the past three years been conducted consistently with an expansionist bias, even though the expansion has been accompanied by rising (though in real terms still negative) interest rates. The wherewithal to cover financial strain on companies caused by the time-gap between conceding higher wages and receiving higher prices has been financed by the banks. One has only to look at the massive overdrafts of companies like British Leyland and many of their companions in the labour-intensive and blackmail-prone engineering industry to realise how the banking system has produced the finance that has in part contributed to wage inflation.

When companies get beyond the help of commercial banks the coffers of the Exchequer open wide: to wit, the Harland and Wolff yard, Upper Clyde Shipbuilders, Rolls-Royce, ICL, BSA and now Leyland. This propping-up happens even more directly and dramatically within nationalised industries. The widening deficits of the National Coal Board and of British Rail are self-evidently due to exorbitant wage concessions which for the time are being financed by larger Exchequer contributions. The Government has requested, and even instructed, the nationalised industries to make ends meet by charging commercial prices. So far these are words. Actions have

[71]

yet to follow and, meanwhile, the borrowing requirement of the public sector, the main fount of inflation, goes on increasing: from £2,700 million, the figure estimated in the March 1974 Budget, to £6,300 million to which it was adjusted after the brief interval that led to the November 1974 Budget.

These additional Exchequer outgoings are not being financed, and cannot be financed, by mobilising genuine savings and by sales of medium- or long-dated Government securities to the non-bank private sector. Instead, the widening Exchequer deficit is in large part being covered by short-term borrowing of Arab oil money. This is the most volatile of bad money. The accumulation of this kind of debt lays up a heap of trouble. For the rest, an undue part of the borrowing requirement is being financed by short-term domestic borrowing, mainly by Treasury bills. These in turn provide a larger basis of eligible liquid assets on which the commercial banks can expand credit by dint of a multiplier which under the prevailing ordinance of credit control is now eight-fold and not $3\frac{1}{2}$-fold as it was when the banks were called upon to maintain a 28 per cent liquidity ratio.

The size of the deficit or borrowing requirement, combined with the abject state of the gilt-edged market, provides most of the explanation for the excessive increase in the money supply; but it is an increase consequential on the policy of trying to push up the rate of growth in the economy, and consequential also on the widening of the financial deficit of the public sector in which higher wages, unmatched by higher productivity, have played a major part.

Measures to cure inflation: courage and discipline
The Government and the Bank of England could in theory stand firm and refuse to expand the money supply. But to do this they would have to make the appropriate changes in monetary and fiscal policies. To turn off or even turn down the money supply tap is not a mechanical, technical operation. It will be the resultant of measures with explosively political content. It calls for a reduction in the Budget deficit, preferably by cuts in expenditure but if necessary by increases in taxation. It will demand a continuing discipline of severe credit policies. It will require the courage to face up to an increase in short-term unemployment and to exorbitant wage demands. To suggest that all that is needed to cure inflation is to

[72]

turn off the money supply tap is to shirk the real problem and fail to consider the social and political forces and compulsions that have caused this money supply tap to gush.

The Heath Government tried to go beyond these monetary simplicities by preaching and at first acting on the principle of a more self-reliant society, allowing market forces to have their sway, providing higher rewards for success and penalties for failure. Alas, they quickly beat a retreat in face of a modest rise in unemployment. Credit policy was eased with the introduction of *Competition and Credit Control*. A massive increase in the Budget deficit was consciously sought. A trail of rescued and well-cossetted 'lame ducks', and dismal surrenders to extortionate wage demands, marked the road of their inflationary retreat.

Their successors have done no better. Their reliance on the Social Contract has been selective and has had next to no impact on the forces of wage-induced inflation. The Budget deficit rose apace and with it the debauch of the currency continues.

Lack of leadership

Why have recent governments, Conservative and Labour, failed in this respect? One reason is that there has been no one on either front bench who has commanded the eloquence and power of logic and conviction, the qualities of leadership, that will carry the case for honest money to the people. And so the field remains dominated by the articulateness and emotive strength of the militants and of the expansionists. The militants will see to it that exorbitant wage demands are tabled and satisfied. The expansionists (and this euphemism for inflationists, floaters and devaluationists includes Ministers, civil servants, academics and journalists) will see to it that monetary policy, the money supply and the rate of exchange obediently adjust themselves to the impulses of wage-induced, cost-push inflation.

Let us not, therefore, fall for the temptation to put the whole blame for inflation on that convenient, impersonal scapegoat, the money supply. A stricter control of the money supply must, of course, be an essential concomitant of any anti-inflationary policy. If the Government did not continually expand the supply of money, the more basic forces making for inflation could not operate so easily. But I emphasise that the success of this stricter control of money

[73]

supply depends on other things being done: first, on some reduction in the public sector deficit now being financed by inflationary means; secondly, on some diminution in the escalation of wages and salaries which has played a major part in creating this inflation-begetting deficit.

We must look beyond the figures of money supply to the social and political forces that have reduced monetary policy to the role of an obedient and even cowardly servant. And the most strident, baying orders of command, reducing that servant to a quivering jelly, have come from militant, well-organised trade unionists with their successful demands for inflationary wage increases.

Prescription for monetary discipline
What then should be done to restore monetary discipline?

First, reduce the Budget deficit. Among the desirable cuts would be the elimination of food subsidies which represent a minute element in the cost of living but a much larger item in government expenditure. A verbal start on this journey has been made by the Chancellor's instruction to the nationalised industries to charge economic prices.

Secondly, restore profitability and with it confidence to the private sector by easing and ultimately disbanding price controls. A promising start was made here too in the November 1974 Budget and in unfreezing commercial rents.

Thirdly, maintain a rigorous credit policy, allowing interest rates to rise to be positive in real terms. Once inflation was seen to be on the wane these rates would rapidly fall well below the present levels.

Fourthly, imbue the Government – of whatever uni- or multi-party composition – with the courage to stand up to exorbitant wage demands in the public sector (which has led the wages-prices spiral), and not to panic in the face of a modest corrective increase in unemployment. Given mobility, retraining facilities and adequate redundancy and unemployment benefits, the corrective will redound to the benefit of all.

[74]

Too Little and Too Late?

MALCOLM R. FISHER

Lecturer in Economics,
University of Cambridge

Rapid inflation and rising unemployment affect most trading nations at the dawn of the New Year. Yet for policy guidance in Britain, it is vital that the similarities should not be allowed to obscure the differences. In what follows we look first at the features that trading nations have in common at this time, and only then superimpose our special features on the analysis.

Act 1, Scene 1: The World, December 1974

The economy of the Western world is in large-scale re-adjustment. Member-states have just passed through simultaneous peaks in expansion caused by, or coincident with, the imposition of a cartel pricing policy upon one of the principal factors of that expansion, oil. Successful cartelisation has led to a vast inter-country redistribution of income, shifting resources to primary-producing oil nations of limited industrial base and scant population. In a sense, Saudi Arabia, at least, is trading oil against rights in industrial assets, and in gold. The outcome for many oil-consuming nations is an immediate adverse shift in terms of trade coupled with a substantial transfer payment to oil-producing nations. Balances of payment of oil-consuming nations must now be balanced more through favourable swings on capital account to offset less favourable balances on current account. Oil-producer nations have higher claims on oil-consuming nations' resources both currently and into the indefinite future. Britain and the West can alleviate these pressures only by acting as a group of consuming nations, not by reckless trade wars which shift the burden from one to the other, merely divert resources, and lower the national income of the group as a whole.

In domestic trade, the required shift in resources must be towards fuel-economising industries, and towards the long-run, less expen-

[75]

sive, alternative fuels, which in the period of cheaper oil were at or below the remunerative margin of production. The disentangling and regrouping of resources is more easily accomplished in countries with some slack in resources and/or a high degree of flexibility. Industrial expansion has, for so long, been built upon falling cost of energy and transport that it is not easy to face up to a reversal. Already, decline in world trade is providing the elbow room for these adjustments and they will take place provided the declines in total demand do not trigger off substantial depression. Unfortunately they probably will.

New patterns of (falling) world trade

World trade is shrinking and this will work to the relative disadvantage of the large oil importers, and those for whom exports are orientated towards manufactured goods in which transport costs loom large in relation to their final value, or are heavy fuel-using in production or in eventual consumption. It is difficult, at the moment, to foresee the required degree of switching from traditional patterns of production and trade.

Necessity is the mother of invention, and new, unsuspected, cheap sources of energy may be forthcoming, but not immediately.

While it is important that each country does not shift deflation on to the others, so that the pattern becomes endemic, trade deficit countries must deflate relatively to trade surplus countries. Those countries which show the most resilience in these matters will be the first to restore more acceptable standards of living and growth.

The so-called recycling of oil receipts, an issue which gets so much publicity, depends upon both oil producers and oil consumers. For the producers, at least, chosen patterns of recycling will be judged in terms of likely patterns of resilience; inefficient and inflexible nations will be progressively shunned.

Act 1, Scene 2: Britain. A perspective, especially after World War II In this context, where does Britain stand? Her obsession with 'security', by which I mean concern with social measures for reducing hazards arising from unemployment, poverty, illness, etc., has led to the relative denigration of economic progress, without which real security is impossible. At the end of 1974, she finds herself a debilitated state, largely through self-inflicted wounds. All sorts of excuses can be offered, but none is satisfactory. Britain was the first country to benefit from industrial revolution, but some who followed her have virtually had two 'break-throughs' to her one. Admittedly she fought two vast wars to the sapping of her strength in manpower and resources, but even those less successful in war have twice built up great industrial strength since 1918. Britain was relatively stagnant between the wars; compared with Western Europe her performance since 1947 has been meagre, though comparing favourably with that of the years before 1914.

'Full employment at any cost', the battle-cry of the Heath administration, typifies the prevailing mood. The Great Depression is still being fought today. In a sense we could say that the inter-war years, as a whole, were times of disappointing economic performance. It was not merely the concentrated unemployment of the 'thirties that was a cause for concern. Yet the fear of unemployment has struck home and led to the underlying succession of actions which can be summarised as interventions to maintain levels of employment at all times by systems of 'fine-tuning' derived as policy conclusions from Keynes's explanation of depression. In England, even today, the Keynes of the particularly narrow interpretation of *The General Theory* is the only economist that matters. This appears to be true of the Treasury and of 95 per cent of the economics profession.

The neurosis of full employment
Growth, specialisation and development of markets in new avenues and techniques are at least of equal importance, yet at all times maintenance of employment has been given precedence. If a firm or industry falters, this is thought bad, for employment may be threatened – hence it must be shored-up by subsidy or other protection. Keynes's policy for depression – that it is better to dig holes in the ground for remuneration than do nothing – has been carried forward to more favourable times in the form: it is better to support low- and declining-productivity industries than foster those with high, and expanding, productivity. Liberal compensations have been forthcoming for the unemployed and the redundant. Since costs of adjustment cannot be avoided, they have been borne elsewhere by slower growth in production through denigration of profit-making

(here success is judged disreputable) and by taxation; thus these incentives have been dampened.

The notion that income tax will, at the margin, through its 'income' effect, induce people to work harder, has bitten very deep. No thought has been given to the possible substitution of work abroad by the able and enterprising under such restrictions, though now the Chancellor is trying to close this outlet. Once again equal treatment amongst British residents is given precedence over progress. This example may seem unimportant, but it is indicative of national attitudes. Moreover, the sum of a series of actions of this type is no longer negligible.

It is important to look at these developments in a long-run context. If Britain wants 'security' she can have it at a cost, and the more 'security' she wants, the higher the cost. By her chosen route that cost is going to take the form of slower rises in living standards than those in countries with more progressive goals.

It would seem that, in Britain, many consider these costs worth paying – at least, the attitudes of all political parties suggest it. If this is granted, we must cut the garment according to the cloth. We must reduce dependence on imports – the list of which covers many of our traditional, most needed, goods and factors of production. Some see alleviation through North Sea oil. By prevailing standards we know it is a high-cost source of energy. We do not yet know that it will provide a substantial flow at competitive prices in the future, and it may be a very expensive way of reducing the import bill.

Britain is a high-consumption, low-investment economy; this makes it difficult for her to take deflationary action without creating internal resentment. She appears to work under the constraint that adjustments must on no account be borne by unemployment. To achieve this might have been possible with adequate foreign exchange reserves, and/or a willingness to let the exchange rate fall, but again her version of 'full employment at any cost' gets her into difficulties.

Act 1, Scene 3: Britain. Today

From 1971 mounting unemployment was tackled by deficit financing on a massive and unprecedented scale. The costs were borne by printing money, and by mortgaging our future income through substantial foreign borrowing. In 1972 the constraint of a fixed exchange rate was removed – a sensible policy in itself, but dubious if merely done to pursue unrestrained monetary and fiscal expansion. No advocate of flexible exchange rates has ever argued that the level of domestic demand then becomes immaterial. Evidently our Treasury experts knew better.

By late 1972 other bottlenecks were encountered as supplies of certain resources were depleted, and neither time nor slack elsewhere was available to ease them. Prices and wage levels inevitably rose through the demand *putsch*, but government, then as now, blamed everyone but itself for the developments. Moreover, by laying blame on sectional interests, notably the trade unions, it created a degree of militancy which it now finds hard to combat. This considerable and vocal group now demands both rising money wages and continuous employment, a combination that can only be delivered, if at all, by a decline in real purchasing power, and, given a low level of resilience and adaptability, by lower relative levels of living in the future than in other advanced countries. Turning the printing press of money supply to assuage the trade unions is not going to give them satisfaction. As activity adjusts itself to higher rates of inflation thereby induced, unemployment will recur, prompting government to increase the dose of the same inappropriate medicine. Whatever other damage they do, trade unions have little to do with the production of inflation.

The effects of unrestrained consumption

Unrestrained home demand has brought with it a colossal import bill. At the same time, disproportionate price rises at home have been undermining our competitiveness in world markets. Even where prices have been held, there have been substantial costs to our export trade by contracts lost through poor deliveries, and by declines in quality of finish and after-sales servicing.

Britain has an unsustainable position in its balance of payments. To put it bluntly, we can no longer assume that our economy is so vital to other countries that it must be shored up because we are tacitly stipulating that all adjustments to the world economic environment must be borne by others with no cost to ourselves in employment. Unlike the Arabs we have no monopoly strength to accomplish this feat. There is no escape from a reduction in our real

[79]

standard of living in the short term. It is as yet unclear how that fall is to be distributed amongst the populace though, from past experience, there is no way that does not involve substantial unemployment over a period.

At least until very recently, and probably now also, relevant domestic action for easing the adjustment process has not been taken, and those measures introduced have exacerbated the situation. We seem to have reached the point where we are so frightened to make any meaningful adjustment that we shall be forced to do so through an exchange crisis triggered off by a serious loss of confidence by foreign holders of sterling. If they had more confidence in our resilience, and powers of recovery, they would start to buy our equity shares at prevailing knock-down prices!

The costs of delay

For Britain, even without the energy crisis, there is no escape from a reigning-in of home demand to a volume more commensurate with our capacity. With the inflation rates running here even before 1974, and *a fortiori* those running now, this adjustment will inevitably be prolonged. Precipitate action on demand could even produce worse results than the disease. Yet, as 1975 dawns, one must wonder whether the sensible option – *gradual, and sustained, reduction in monetary demand at current prices, and hence in the domestic money and overseas borrowings used to finance it* – remains open. The costs of delay are already far too high. Further delay could endanger not merely levels of living but the whole fabric of our democratic way of life through loss of confidence in money, that supremely successful aid to growth of markets, when properly harnessed. If other trading nations are successful in damping inflationary price rises, as many seem to have done, our relative rates of inflation will glaringly reveal the danger of our position.

Britain then faces the general world problems of a regrouping of factors through the uncompensated rise in a basic and pervasive source of energy coupled with the contraction in world trading. On these, it has superimposed its own internal problem – an attempt to generate home demand and sustain activity beyond the capacity of the economy. The transitory benefits of this policy, if not already apparent in late 1973, would shortly have been so. In 1974 these problems became entangled: at the end of the year Britain suffered

[80]

unprecedented rates of inflation and Government Borrowing Requirement, whilst running a stupendous deficit on foreign account. This combination cannot be maintained. The only operative question seems to be whether we are going to institute corrective action of our own volition, or wait to be pushed into it.

Short-run alternatives – and dilemmas
At first glance it would seem that the curtailment of demand for our exports will provide the margin of capacity to permit industries that have grown up under *excessive* pressure of home demand to survive. Perhaps this explains the Government's reluctance to restrain demand. It may hope that the temporary employment effects it engineered may be prolonged. Such confidence is not justified, for we must reckon with the foreign payments deficit, already very large, and likely to swell as export demand declines faster than demand for imports. With exchange reserves very limited, Britain must rely on its attraction as a placement area for oil revenues of Arab States, an aspect of doubtful comfort in one of the less resilient economies, or allow the exchange rate to float freely, and that means downwards, after the interventionist support of recent times.

Yet we cannot be sure that the short-run price elasticity of demand for exports secured will offset the downward shift in the demand for exports. Moreover, recent bottlenecks in supply must make it very doubtful whether industry can speedily obtain the labour and capital needed for additional export supply. With little short-term improvement in the balance of payments, there must be retrenchment of public expenditure, higher taxation, or direct control over imports. Control over non-oil imports would impede the industrial adjustments needed, while control over food imports would fully expose the flimsiness of the so-called 'social contract' as their prices soar.

In setting out these alternatives, we have ignored the harm done by policy measures of the recent past – harm that urgently needs to be undone. We mention but three, all of which have borne especially severely on the business sector: advance corporation tax requirements of the March budget, main control measures of Phase II and Phase III of an incomes policy which contained no safeguards for the excessive rates of inflation since experienced, and the sharp monetary squeeze of early 1974. A business sector that, through no

[81]

fault of its own, is fighting to save its life is not likely to be adapting itself to the new energy-expensive world. Apart from specific cases of hardship created in this way, any policies of relative alleviation to the business sector should be general, rather than specific. We have not the information to know whether a particular company should be shored up. Moreover, protection of jobs in redundant industries, by subsidisation, is going to do nothing to increase adaptability. It will only result in aid for the less progressive industries, and this works to the disadvantage of the more progressive. It is better to let those displaced seek more long-run work opportunities forthwith.

Obsession with redundancy avoidance carries with it the implication that existing firms with their conventional ranges of products and factor mixes will be as relevant in the world of tomorrow as they are today. That cannot be so. For labour, the policy does more harm than good. If people are to train for fresh tasks at a cost to them both in goods and in time, they need as long a period as possible in the new activity if they are to secure an adequate return on their retraining investment.

Act 2, Scene 1: Britain. The near future
There is little evidence, so far, of any trimming of home demand on our resources. There has been some curtailment of the money supply for short periods, but financing the budget deficit seems then to have been shifted to overseas loans at undisclosed rates of interest. There has been much intervention to hold the exchange rate of the pound. The internal budget deficit and foreign deficit make it impossible to maintain the present position. For the moment we seem to be, like Mr Micawber, waiting for something favourable to turn up. This seems especially hazardous when we compare our lot with that of other trading nations, all of whom, save for Japan, have markedly lower rates of inflation, and generally more productive and versatile economies. Yet we seem to want to wait until we can mischievously and incorrectly blame other nations for our predicament. We should be asking ourselves whether they will be anxious to bail us out of our troubles, many of which are self-inflicted, when we make no effort to help ourselves.

Policy for self-help
In my opinion a policy of self-help would consist of a curtailment

[82]

of home demand, especially government expenditure at current prices, sustained over quite a long period, unless export demand falls off very heavily. I would try to shift resources on balance towards the business sector to offset the particular penalties already applied. Market prices generally should be allowed to record the changing costs of energy in industrial as well as in consumption sectors, so as to ascertain the more remunerative industrial groupings of the future. Inevitably consumption will be cut: that must happen anyway.

Unemployment will inevitably rise through this action, but less, I would submit, than under any alternative policy.

The more likely 'scenario' is different. A crisis will develop through the balance of payments. There will be a run on the pound, and a sharp fall in the exchange rate. As the 'J'-curve effects[1] are all too likely to predominate, the begging bowl will have to be produced again, but the terms for any contributions will be onerous. The necessary corrective measures will still be the same in terms of domestic expenditures, but more stringent and long-lasting because the conditions for foreign aid will be much more severe.

My policy advice is to curtail demand while there is time. My expectation is that we shall act, if at all, too little and too late, and we shall have only ourselves to blame. Hence my gloom on the eve of 1975.

Make no mistake: we are already in an inflationary depression. Those nations that most speedily adapt themselves, and learn from these adverse circumstances, will secure the relative advantages of the recovery phase.

---

[1] ['J' curve effects: describes the sequence of the impact on exports and imports of an effective devaluation of the £ – first, an adverse effect from the immediate rise in import prices and fall in export prices, followed in time by an expected more than offsetting improvement from the consequent reduction in import and increase in export volume. – ED.]

The Complaisant Economy
RALPH HARRIS

Lecturer in Political Economy, University of St. Andrews, 1949-56
General Director, The Institute of Economic Affairs

IF POLITICIANS of both main parties are blamed for the record of post-war economic policies, what are we to say if they reply that it was mainly the fault of economists who encouraged them to pursue at least half a dozen objectives now seen to be mutually inconsistent ? Full employment, stable prices, increased welfare, higher consumption, favourable balance of payments, fixed parities, low interest rates – all were promised simultaneously and every one is now forfeit or threatened along with a curtailment of economic freedoms.

Confusion of counsels

Politicians naturally wish to believe that all their hopes can be fulfilled – if only their party is given power. Driven by the superficial mathematics of democracy under at most five-yearly elections, they try to maximise short-term popularity and are understandably tempted to rule out unwelcome choices as 'politically impossible'. Economists are under no such compulsion. Even as civil servant advisers to governments, their life tenure should enable them to emphasise the economic realities and the limited ability of governments to soften them. Professional economists may differ among themselves because of divergent philosophical values or of disagreements on technical analysis. But confusion of counsels caused by philosophic differences can be reduced by candid declaration and of technical differences by more regard to empirical evidence, of which post-war experience has provided a daunting abundance.

Since 1945 a large source of conflict in economists' testimony has been due to recommendations framed not in the light of their admittedly narrow but indispensable expertise, but in the twilight of their view about what is 'politically acceptable'.[1] The result has been to

---

[1] '. . . people who formulate the structure of the models on which decisions are taken try to incorporate what they call "political realism" . . . The result . . . is bad economic theory and bad political theory.' (Samuel Brittan in *Inflation: Causes, Consequences, Cures*, IEA Readings No. 14, IEA, December 1974.)

impair their prestige and influence. Not only are economists untrained in judging what may be 'politically possible'; their authoritative judgement about the economic component in a given situation should, on the contrary, play a significant, often dominant, part in shaping what politicians (and commentators and voters themselves) come to regard as 'practicable'.

Alas, as Pigou warned 40 years ago,[1]

'the ambition to play a part in great affairs . . . to stand near the centre of action' may lead 'a young man . . . to make slight adjustments in his economic view so that it shall conform to the policy of one political party or another'.

In the post-war years, the larger role which politicians of all parties have claimed for government intervention in economic affairs has led to a more debilitating, less obvious, danger. Instead of an open clash of economic debate between partisans of alternative views on the limits of political control, there developed a 'consensus' which ruled economic prescriptions out of court as 'politically impracticable'.

The resulting dilemma has not been better expressed than by Sir Dennis Robertson in his Presidential address to the Royal Economic Society which 25 years later bears an almost uncanny ring of truth.[2]

' . . . it takes some spirit to state clearly and fairly the case for wage reduction as a cure for unemployment or an adverse balance of payments, or the case for the curtailment of subsidies and the overhauling of the social services as a solvent for inflationary pressure, without being prematurely silenced by the argument that nowadays the trade unions would never stand for such things. Perhaps they wouldn't; but that is no reason for not following the argument whithersoever it leads.'

That advice applies no less today. The longer it is ignored the worse our economic condition. Had it been taken in the 1960s we should not now be facing an economic collapse that has led historians and political observers to see a breakdown in democratic institutions as itself politically possible.

Full employment and inflation

It may just have been excusable that, in the post-war impulse to

---

[1] 'An Economist's Apologia', reprinted in *Economists in Practice*, Macmillan, 1935.

[2] 'On Sticking to One's Last', reprinted in the *Economic Journal*, 1949.

remedy all the supposed grievances of history, economists should have hesitated to question the mixture of idealism and opportunism which comprised the dominant political philosophy that has come to influence all three parties. Thus Keynes had provided a theoretical foundation for the belief that *mass* unemployment could be corrected by deliberate government action to raise the level of aggregate demand. Yet by 1949 the first devaluation of sterling had demonstrated the practical difficulty of maintaining a high enough pressure of demand to hit the ambitious target of 'full employment' without pushing up prices, including wages as the price of labour. Disappointment did not, however, lead to a revision of policy objectives or a more modest definition of 'full employment' which Keynes had pitched well below the level post-war governments set themselves.[1]

Instead of revising the level of unemployment that could be regarded as 'politically acceptable', politicians were encouraged to persist with a succession of expedients which might curb the inflationary consequences, starting with the first 'incomes policy' of Sir Stafford Cripps in 1949. Evidence nevertheless accumulated that the tightrope act of successive Labour and Conservative Chancellors to tiptoe between unemployment and inflation was accompanied by alternations between credit expansion and squeezes which disrupted the long-term investment required to match capacity to the steady expansion of demand for consumer goods and services. The kindly imagery of 'accelerator and brake' gave way in the 1950s to denunciation of 'stop-go'. But so long as more realistic policy objectives were ruled out as 'politically unacceptable', more and more economists were tempted to offer confident panaceas that may have been 'acceptable' but were unproved and swiftly shown ineffective.

Taking their cue from politicians, some economists too readily joined the uncritical crusade for 'growth'.[2] By the early 1960s it passed as 'sophisticated' to believe that the increasing demand necessary to sustain over-full employment would not inflate prices so long as increasing supplies of goods and services could be conjured out

---

[1] After 1945 Chancellors of the Exchequer thought 'reflation' appropriate whenever unemployment exceeded 300,000 where Keynes had considered 700,000 might be normal in a labour force exceeding 20 million. (Lord Kahn, 'What Keynes Really Said', *Sunday Telegraph*, 22 September, 1974.)

[2] One of the few who withstood the tide was Colin Clark: *Growthmanship*, Hobart Paper 10, IEA, 1961.

of the hat to absorb the swollen purchasing power. It seemed that economic jargon and naive macro-statistics were a positive barrier to the promptings of common sense. Asked how perpetual growth was to be generated so as to keep ahead of an inflationary pressure of demand, Mr Selwyn Lloyd gave us the National Economic Development Office. When that failed to square the circle, Mr George Brown gave birth to the National Plan[1] which predictably paved the way to the second devaluation of sterling in 1967, despite a succession of 'incomes policies' which, failing to stem the tide of inflated demand under Mr Harold Wilson, were duly intensified by Mr Edward Heath with equally little lasting effect.

Neglect of scarcity

'Growth' was also pledged in advance to meet 'rising expectations' for increased personal standards of living on top of the continuing increase in collective consumption by ever-expanding social welfare. Governments which chastised private extravagance indulged the fallacy of 'spend now in the hope of being able to pay later'. Taxation of profits and personal incomes was pushed to levels that cut into the funds available for private investment and saving which were at the same time expected to provide for the renewal and extension of capital on which growth was thought to depend. When government revenue proved insufficient to redeem electoral pledges of increased benefits, government ran budget deficits defended as keeping the demand for labour at 'acceptable' levels.[2]

It was not only in monetary policy that politicians were complaisant. From modest levels under the first Labour Government, spending on health, education, housing, pensions and related social benefits grew to account for almost half of total public expenditure, which in turn has come to pre-empt more than half the national income.[3] Still bemused by the prospects of 'growth', too many

[1] The IEA published a warning against the exaggerated hopes of *The National Plan* in an Eaton Paper (No. 4) with that title written by John Brunner in 1965 (3rd Edition, 1969).

[2] In *How Much Unemployment?* (Research Monograph 28, IEA, 1972), John Wood exposed the inadequacy of official statistics on which the Conservative Government was then basing its economic strategy.

[3] There is no space here to list the dozens of *Papers* the IEA has published since 1959 on the shortcomings and consequences of government domination of the markets for education, medical care, housing, pensions. Much of this work is summarised in *Choice in Welfare 1970* (IEA, 1971).

economists (and economic institutes) refrained from cautioning politicians that governments could not take more cake for themselves whilst urging that more should be invested, exported, given to developing countries and every other deserving cause.

At its simplest, the root cause of the grave economic reckoning now threatening to overwhelm us in 1975 stems from the neglect of the first-year student's concept that since resources are scarce the 'opportunity cost' of any claim on them is the sacrifice of alternative uses to which they could be put. Had more economists 'stuck to their last' and persisted in hammering home such unwelcome truths, it could hardly have become 'politically possible' for governments to win credibility for policies that have by 1975 set up monetary demands approaching £110 for every £100 of domestic output. Recent policies have not flirted with inflation; they have provided a cast-iron formula for an explosion of wages and consumer prices.

Unions versus money

Much apparent dispute about the 'cause' of inflation is a war of words rather than a conflict of analysis. All the evidence of inflation in many countries over many centuries demonstrates that prices cannot continue rising without an accommodating increase in the volume of money in circulation. If the growth in monetary demand is kept within the rise in real output, few economists would now deny that inflation can be stopped. The trouble is that the demand for labour at current (inflated) wages would fall below the 'full employment' level if the demand for its products were not maintained by accelerating inflation. Most analysts who blame trade union demands for inflation seem to accept that labour would price itself out of jobs by insisting on excessive wage increases; but they go on to assume the resulting unemployment would be so politically unacceptable that governments are forced to inflate monetary demand to match the rise in costs. Thus the dispute between 'monetarists' and exponents of 'cost-push' can be reconciled by the formula that the unions are the active 'cause' of inflationary pressures so long as the government obliges by enlarging the supply of money to maintain 'full employment' at whatever wage level unions and employers care to agree.

In effect, therefore, governments underwrite wage (or other cost) inflation by their complaisant monetary policy. Consequently, union leaders can impose inflationary wage settlements on employers without having to worry about the unemployment that would result in the absence of a permissive monetary policy to maintain 'full employment'. The conclusion, which significantly is winning adherents among previously sceptical economists and politicians, is that monetary discipline holds the key to checking inflation.[1] Some still hesitate to declare the truth plainly since, if the monetary screw is tightened, however gradually, while powerful union leaders think they can go on their old inflationary ways, the cost in terms of unemployment must be more severe. Hence the lingering attachment to 'incomes policy' to hold back wages so that the brake can be applied to monetary expansion with the minimum side-effects on employment.

The danger for economists of devising ever more ingenious incomes policies is that they thereby direct attention away from the essential, but unpalatable, pre-condition, namely that total monetary demand must be reduced towards the level consistent with the value of domestic output at uninflated prices. Economists should leave politicians to judge the 'practicability' of incomes policy and meanwhile push their own analysis further. It is not the competitive market pressure of unions that enables them to dictate the wage employers must pay them; it is the coercive power of their entrenched monopoly position in many industries to withhold labour and alternative sources of supply with the help of picketing, 'blacking', social security benefits, tax repayments and the rest. Economic analysis provides no authority for urging an incomes policy on politicians when the optimum solution would be to remove the monopoly power by law and judicial review as has long been enforced against the less damaging restrictive practices of private companies (though not yet of nationalised industries).

Where now?

As an economist who puts a high value on freedom of choice for man as consumer and producer, I should wish to see the necessary monetary cure for inflation supplemented by wide-ranging policies to improve the mobility of labour between industries, firms and localities. Since housing and job demarcation are severe obstacles to wider choice of better-paid employment, we should phase-out controlled

[1] The IEA's effort to question the post-war neglect of monetary policy started in 1960 with *Not Unanimous: A Rival Verdict to Radcliffe's on Money*.

and subsidised rents and phase-in a massive programme for re-training redundant workers – whose numbers would be vastly increased by removing the over-manning in government as well as in private employments that have been bloated and shielded by 30 years of mounting inflation. Such long-run institutional reforms would also bring a better promise of real growth than the shifting expedients favoured by Labour and Conservative governments.

Having seen the costly folly of short-term forecasting, even in the confident hands of the NIESR, I am reluctant to offer a firm prediction about 1975. But there can be no doubt that the strategy of Mr Healey places the British economy at the mercy of events abroad over which we can have little influence. Whilst claiming to curb the monetary excesses of his Conservative predecessor, his last Budget (complaisantly) accepted a rising real level of public spending running more than £6,000 million in excess of public revenue. He pinned his hope of preventing such a deficit from leading to an inflationary increase in the domestic monetary supply on the alternative of borrowing from abroad. In round terms, the gap between domestic claims on resources amounting to around £76,000 million and the supply of domestic output of around £70,000 million will be filled by importing that much more than we export (apart from any domestic private saving that may prove possible).

While something like 8 per cent of our current standard of living is thus provided by courtesy of foreign lenders, the British economy must remain vulnerable to the sudden removal of these overseas props. Certainly, if the outlook for continued inflation is as gloomy as many of us see it, the chance of a sudden withdrawal of funds remains ever-present. The foreign exchange value of the pound must then fall further and the living standard we are giving ourselves would be forcibly cut back. Since government expenditure now accounts for more than 50 per cent of the GNP, it cannot much longer remain immune from drastic reduction if we are to live within our national income. Accordingly, the monetarist prescription of a cut in the budget deficit should be implemented by a substantial reduction (at least 5 per cent) in central and local government spending. This should be seen as the first instalment in a continuing programme of public economy that will need to proceed all the faster if we are to reduce corporate taxation so as to restore profitability and to free resources for company investment. Less costly government intervention and less subsidy throughout industry, in social welfare, on price controls and the rest would release energies, stimulate efficiency and widen freedom of choice. Who still doubts that this course offers a better recipe for increased production than the post-war policies that were thought 'politically acceptable' but have proved economically disruptive?

A more modest role for imperfect government
If the impending collapse compels such a redirection of public policy, then whenever any well-meaning lobbyist urges that our difficulties can best be met by turning from markets to governments as agents of economic progress, the consensus reply is likely to echo the words of Alfred Marshall: 'Do you mean Government all wise, all just, all powerful, or Government as it now is?' As we tread the minefield of 1975, economists should recommend for government a more modest role in our economic affairs to match its inbuilt imperfections.

Turning Point or Moment of Danger
E. VICTOR MORGAN

Professor of Economics,
University of Reading

THE *Concise Oxford Dictionary* gives two meanings for the word 'crisis' – first, 'turning-point, especially of disease', and, secondly, 'moment of danger or suspense in politics, commerce, etc'. In the latter sense of the word, we have passed through innumerable crises in the past 30 years, and will experience many more in 1975. In the former sense we have, unfortunately, not had a crisis at all, and 1975 is most unlikely to bring one.

Moments of danger or suspense

'Moments of danger or suspense' will arise for many reasons. Production will be stagnant or falling and unemployment will rise, though probably not very much. In the private sector a number of firms will either go bankrupt or have to be rescued with an infusion of cash from other firms or from the government. In the public sector we shall pay a lot more to the miners without getting much more coal, and we shall give smaller increases to teachers, dustmen, postmen, transport workers and nurses with the result that standards in these branches of the public service will deteriorate still further. The rate of increase of consumer prices may fall slightly at the beginning of the year but will then accelerate again to at least the 1974 figure. The balance of payments will probably deteriorate even further, and in spite of enormous overseas borrowing a further decline in the value of sterling is likely. These events have already been largely discounted on the stock market, but there will be no lack of suspense there, and a lot of money will also be made and lost in the markets for gold and works of art.

Another safe prediction is that many people will lump all these evils together under the name of 'inflation' and will assert that the only cure is a statutory incomes policy, preferably administered by a 'government of national unity'. They may even get their way but they will be quite wrong. Inflation is only a part of our present difficulties, and it is not caused either by trade unions or by oil sheiks.

Power and acquisitiveness

So long as men use power to acquire riches those whose power is growing will get richer relatively to those whose power is declining. In some societies the relevant power is political or military; in our own it is still mainly the power of the market. The major constituent of market power is simply the desire of other people for the goods and services that we own or can produce. In certain circumstances power derived from this source can be enhanced and abused by monopolistic combinations such as producers' cartels and trade unions, but we should not exaggerate their influence. OPEC has only restored the price of oil, in terms of manufactures, to about what it was 25 years ago; the power of the NUM stems mainly from this rise in oil prices; and even the likely exactions of the miners in 1975 will probably leave them no better off than some other workers (e.g. in motor manufacture and newspaper printing) doing less disagreeable and less dangerous jobs.

Whether they come from increased demand or enhanced monopoly power these relative rises in incomes, and the price changes that go with them, are not causes of inflation. On the contrary, if money income in the community as a whole remained constant, the fact that some prices and incomes rose would imply that others must fall. The rise in oil and coal prices would be *deflationary* for the rest of the economy. This point is generally recognised when the government brings about a rise in prices by increasing commodity taxes.

Inflation is a monetary phenomenon

The source of inflation is not to be found in price rises for individual goods or groups of workers but in a general rise in money incomes at a rate faster than the growth of real output. One of the very few messages that comes out loud and clear from the economic research of the last 20 years is that such general increases in prices and incomes are monetary phenomena associated with an increasing supply of money and low real rates of interest.

The crucial question then becomes: how do increases in the money supply come about? Four possible answers can be given:

[93]

1. Because of new discoveries of the precious metals.

2. Because of a balance of payments deficit by a country (such as the US) whose currency forms part of the monetary reserves of others.

3. Because governments want to spend more than they are able or willing to raise by taxation and by borrowing from non-bank sources and,

4. Because governments believe that they will become un-popular if they allow some prices and incomes to fall in money terms and so deliberately create monetary conditions such that relative changes take place not by a combination of upward and downward movements, but by everyone moving up but some moving faster than others.

It is, of course, reasons three and four that have been important in Britain since the Second World War and are likely to be so in 1975.

Britain's inflation is not 'imported'

At this point we should deal with a further fallacy, that concerning imported inflation. With a régime of fixed exchange rates there are many ways in which inflation spreads from one country to another, and it is difficult for one country to maintain for long an inflation rate very different from the rest of the world. In these circumstances a country such as Germany, whose inflation rate was lagging behind those of its neighbours, could fairly complain of imported inflation. Such complaints are inappropriate in Britain for two reasons. First, British inflation rates have been consistently higher than those of most other industrial countries in Europe and North America. Secondly, the world system of fixed exchange rates set up at Bretton Woods began to disintegrate in 1967 and since June 1972 the sterling rate has been floating.

Since we cannot take refuge in the good old British custom of blaming the foreigner, we must ask why our governments are more addicted than others to inflationary monetary and fiscal policies? One possible reason is the widespread belief, with very little sup-porting evidence, that the control of inflation by monetary action must necessarily involve heavy and prolonged unemployment. Another is the equally unsupported belief that trade unions are more likely to be roused to militancy by falling money wages with stable prices than they are by stable wages and rising prices.

These beliefs may explain why governments have neutralised what would otherwise have been the deflationary effects of rises in the price of oil and other imported materials. They cannot explain the chronic tendency of governments to generate a borrowing requirement that cannot be met from genuine savings, i.e. from the non-bank private sector.

Inefficiency and inconsistency

This tendency seems to be related to the general inefficiency of our economy combined with the unwillingness of voters to recognise the fact. We are happy to vote for large increases in public expenditure, yet in our private spending and saving decisions we strive for a rate of growth of consumption that is inconsistent with a rapid growth of public spending and a slow growth of production.

When it comes to asking why the economy is performing so badly we have to make the sad confession that, 200 years after Adam Smith, economists still know very little about 'the nature and causes of the wealth of nations'. The following paragraphs belong, therefore, to the realm of speculation and casual empiricism rather than scientific inquiry.

First, we seem to be afflicted with a blind faith in the power of governments to do good that is quite unshakeable by evidence. Successive governments have pledged themselves to promote faster growth, yet we are falling steadily further behind our European neighbours; to cure poverty, yet the relative position of the 'submerged tenth' has not improved at all; to strengthen the balance of payments, yet the deficit is running at an all-time record; to remedy the housing shortage, yet the plight of the homeless is as acute as ever; to control inflation, yet the rate of increase of consumer prices has trebled about every four years since 1960. In spite of these and countless other examples of failure, Conservatives and Liberals demand more intervention almost as loudly as Socialists; and every time a few people die in a fire or an accident, or a tanker spews oil onto a beach, a Minister of the Crown rushes to the spot and initiates a public inquiry.

[95]

Swollen numbers in government service

Besides inducing a reluctance to fend for ourselves, this attitude has combined with the operation of Parkinson's law to produce a considerable growth in the numbers employed in government service. Changes of classification make it impossible to give a continuous series, but the following figures show the percentage increase in the numbers in central and local government service over periods where a common classification was in use.

	%
1955-59	2.2
1959-69	11.5
1969-73	12.4

The absolute increase in numbers is rather less than 300,000 but, in assessing their significance, we should remember that the increase took place over a period in which the manufacturing labour force has been declining, that most of the people concerned are highly trained, and that every additional civil servant makes additional work in the private sector.

The harm done by the diversion of resources to government intervention has been magnified by the contradictions and vacillations of policy. These could be the subject of a book rather than a paragraph and I can give only a few examples. Policies have been vacillating sometimes because of changes in the party in power, sometimes because of changes of view within parties; examples are nationalisation, the pricing policy of the nationalised industries, prices and incomes policies, state pensions, defence expenditure and the Common Market. Governments have generally discriminated against effort, e.g. by high marginal tax rates and the earnings rule for pensioners. They have discriminated against saving by low real interest rates and the penal treatment of 'unearned' incomes. And they have distorted the capital market so as to channel personal savings into the National Savings movement and the building societies rather than into the finance of industrial investment.

Government controls have impeded the efficient use of existing resources and created artificial shortages (most notably in housing). They have also discouraged private provision in a number of important sectors (e.g. rental housing, schools and hospitals), thereby increasing demands on over-strained public services.

Attempts at 'fine-tuning' have generally magnified the cyclical fluctuations they were supposed to smooth, and the number, complexity and inconsistency of government actions has greatly enhanced the uncertainty that private industry has to face in planning its investment.

Deterioration and breakdown?
What will be the consequences of maintaining present policies and what changes are required? The answer to the first question will not be seen in full during 1975, but the logical outcome of present policies is a continued deterioration in our economic performance in relation to our neighbours, an acceleration of inflation, a further deterioration in the balance of payments and a fall in the value of sterling giving a further twist to the inflationary spiral. This is a recipe for hyper-inflation and the breakdown of parliamentary democracy.

To escape these evils we shall have to live through a crisis in the first sense of the word recorded by the Oxford lexicographers: a turning point that will bring a government committed to do much less and to do it more effectively; a rule to ensure a steady expansion of the money supply at a rate in line with the growth of productive capacity; a willingness to maintain that rule in face of minor fluctuations in activity, and to confine public spending to an amount that can be financed consistently with it; a greater use of the market for the allocation of resources; an encouragement of the private sector, alongside the state, to make provision for social welfare; and a readiness to leave private individuals and companies to reap the rewards of effort and efficiency, and to pay the penalties for their opposites. It does not seem very likely that we shall see this kind of crisis in 1975.

In Thrall to Creditors?
ALAN WALTERS
Sir Ernest Cassel Professor of Economics,
University of London

Cause

THE PROXIMATE cause of the crisis and imminent decline of Britain
is the expansion of public spending and money supply that began
under Mr Reginald Maudling, Conservative Chancellor of the Exchequer, in 1963-64 and which continued, with only a short pause in
1969, for ten years. The ostensible objective was to break through to
self-sustaining export-led growth – 'go-go' instead of 'stop-go', full
employment instead of the 'wastes' of idle hands.

Politically such expansionism has been popular. In the short (two-
year) run output, growth and employment increased; the bad effects
of inflation and declining growth and rising unemployment began to
appear only after a long two-year delay, beyond the far horizon of
politicians. The old discipline of the balance of payments was ignored (except for the period 1968-70); as Mr Harold Lever said,
foreign debts were to be regarded as merely strokes of a pen, only
entries in a ledger. Borrowing was painless.

Most professional economists, particularly the Cambridge school[1]
and the prestigious and independent National Institute of Economic
and Social Research, provided powerful intellectual backing for this
policy. Perhaps the main explanation of their support was the persistent fascination with the short-run and the neglect of the longer-
run consequences. In this respect the policy reflected the meritri-
cious political and adminstrative pressures of the day. But there was
also the strong belief that the price system could work only badly
and inequitably under contemporary conditions; hence government
controls and direction were required. Thus any long-run inflationary
effects could be blamed on the lack of bureaucratic controls over
wages and prices. Indeed some may well have appreciated, and in-

---

[1] It should be noted that the 'New' Cambridge school appears to have developed reservations about such 'go-go' policies.

[98]

deed welcomed, the long-run inflationary consequences as a powerful force eroding capitalism and hastening the collectivisation of Britain.

One of the strange features of the 1975 crisis is that although the professional economists and 'experts', both inside and outside the civil service, gave such bad advice, they have not been discredited. On the contrary they go from strength to strength; it is said that the fault lies not with their policy proposals but with foreigners, oil sheiks, and above all in the government not imposing suitably draconian wages-price controls. Controls failed because they were not applied early or rigorously enough.

The great readjustment

But by 1975 we have gone too far. The grand illusion that Mr Anthony Barber, Conservative Chancellor of the Exchequer, conjured up in September 1971 has hovered too long. The vast expansion of public spending and the money supply has continued at a rapid pace. The rate of expansion was only modestly slowed down in 1974 by Mr Denis Healey, Labour Chancellor. And it will almost certainly be speeded up again in 1975, at least until the crash. The great liquidity crises of 1974 have demonstrated once again that the acceleration of the money supply is like a drug – the more that is imposed on the economy to give the artificial stimulus the politicians require the more the economy needs in order to maintain such over-full employment. The first signs of the painful adjustment process have already appeared before the end of 1974: the increased unemployment, imminent bankruptcies, plummeting investment, and accelerating price inflation. And calls for more expansion to fight the slump are loud in the land.

Big bang

But this is no ordinary inflationary contraction. The balloon has been blown up too high; there will be a big bang rather than a soft pop. The main difference is that, by means of Mr Lever's book entries, Britain has been financing an enormous and unprecedented deficit on the current balance of payments – of the order of £4 billion (where billion = 1,000 million) in 1974 and still rising. At least £2.5 billion and perhaps even more than £3 billion was borrowed short from the sheiks. It is easy to see how the financing of the borrowing requirement of the public sector (about £5.5 billion) was achieved without an inordinately large increase in the money supply; we greatly increased our indebtedness to foreigners. This can be illustrated as follows:

Stock of money (M₃) on 1 January, 1974	£32.0 billion

Minimum 'normal' 1971-73 rate of growth of M_3 at 20 per cent	£6.5 billion
Actual growth	£4.0 billion

Residual: additional borrowing from foreigners (Arabs)	£2.5 billion

(The domestic credit expansion (DCE) was at least £8.0 billion – 25 per cent of M_3 – and almost ten times the limit the IMF suggested in 1968-69.)

Reducing the rate of growth of the money supply by running massive deficits and borrowing from foreigners is no way to conquer inflation. It merely prolongs it. Foreign goods flood British markets, and by adding to domestic supplies they keep down the rate of rise in prices. But this does not beat the inflation; it merely serves to defer it. Ultimately – and I believe soon – Britain must pay for her excess imports with real goods, not paper promises. Interest payments mount at a frightening rate and the principal must be repaid. I would judge that many countries have lent to Britain far beyond the bounds of normal prudence – perhaps because, like the *nouveau-riche* sheiks, they followed old patterns of behaviour.

Debts and devaluations
By this massive lending (or, to use the current euphemism, 'recycling') to the UK the oil-rich nations have permitted the bloated British government sector of the economy to increase further its command of the diminishing base of production; and it has allowed the householder to maintain and even increase his consumption. These borrowings from overseas have not been used to increase the industrial capital and productive potential of Britain; on the contrary, over these last few years our industrial capacity has probably *declined*. Thus Britain has not invested these funds profitably to produce goods which will pay the interest charge. The loans are

being frittered away in consumption and public expenditure, while our industrial wealth is eroding away.

The oil-rich nations have done Britain no favour, only lasting harm, in lending her so much; they have delayed the inevitable day of adjustment and reckoning and made the task of recovery much more difficult. (Suppose, for example, Britain had been forced to cut back in 1973 when the foreign deficit increased to some £1.0 billion; would we not have been in a much better state?) It is becoming increasingly clear that Britain will have great difficulty repaying even her present obligations. Her portfolio of foreign assets, once a comforting strength, has been much eroded by the world slump, and their realisation by sale overseas would now incur substantial losses. The current deficit on the balance of payments will continue to grow during the first half of 1975 and the need for loans will also grow. Clearly the borrowing must stop – soon. Funds will be shifted not to London but away from Britain. There does not seem to be any way to avoid a massive flight from sterling.

Cutting our imports by some 20 to 30 per cent in a relatively short period is the immediate problem to be faced. A substantial rise in the sterling price of imported goods can be brought about by a steep devaluation of sterling. This devaluation is likely to be more than that required to make UK exports competitive in world markets and suitably harden the UK market for foreign exporters. The exchange rate will anticipate inflation and the future decline of sterling (as one can see in inflationary Chile at present). It is difficult to judge the level to which sterling would sink against the US dollar – but in a free-fall I would suggest that $1.60 might be appropriate. Such a devaluation would not *cause* inflation. On the contrary, devaluation is simply a *consequence* of the inflationary expansion of public spending and the money stock which has persisted since 1971. Nevertheless the devaluation will be associated with a considerable acceleration in the rate of inflation as the prices of imported foods and raw materials rise. And it is likely that cause and effect will be confused yet again and the devaluation will be blamed for the further inflation.

A siege economy
I suspect therefore that whatever government is in power during 1975 will mistakenly try to cushion the blow. 'Temporary' import

controls – such as quotas on 'inessential' imports – and some rationing of basic foodstuffs, steel and fuel and perhaps other commodities, will be seen as the 'fair' way to share the burden. More stringent controls on foreign currency and capital flows, and maybe even the Treasury acquisition of foreign assets in the hands of UK residents, are likely. Such expedients, together with what residual borrowing we can still extract from an unwilling world community, may serve to prevent the steepest form of depreciation of sterling. The inefficiencies and gross inequities that will be generated by such controls and rationing are well known. (We may recall that it was only after Hugh Gaitskell's 'bonfire of controls' in 1951 that Britain began to emerge from the post-war gloom.)

It is likely that the immediate policy after the crisis breaks will be a wage and salary freeze on percentage increments together with a similar control over prices. The trade unions are (December 1974) now pursuing the only rational and sensible policy in trying to get their increments before the freeze is imposed. Insofar as such action reduces the effectiveness of the freeze, and if it be recognised that when wages rise prices must also rise, it will *help* not hinder the adjustment process. (To avoid misunderstanding I should make it perfectly clear that there will be great inequities and unfairness in such a wage-grab and price-push. But the inequities are a consequence of union power which exists and will be exercised anyway, whatever may be said about the Social Contract, and that power has been and is being much intensified by the inflation itself.) The less effective the wage and price controls the better for Britain. Similarly the greater the fall in sterling the more speedily we shall get out of debt. The gnomes of Zurich will do Britain a major service by talking down the pound; the politicians of Westminster will do much harm by trying to prop it up. All one asks for is the long overdue recognition of reality.

It is impossible to say how long it will take to get Britain back on to a stable-price growth path again. I suspect it will take many years, partly because we have slipped so far, but mainly because I fear that we need to learn all over again the lessons we learned in 1947-51.

The Biggest 'Stop' Year since the War
MICHAEL PARKIN

Professor of Economics,
University of Manchester

PREDICTING future economic performance is always difficult but at the beginning of 1975 it is even more hazardous than usual. We are living through a period of economic history which contrasts sharply with the years of stability and sustained growth of the 'fifties and 'sixties, the most obvious parallels of which are the inter-war years. The British and world economies are currently reacting and responding to the massive change in relative prices inflicted by the oil producers in the autumn of 1973 and to the most sustained and savage period of monetary tightness since the late 'twenties. Exchange rates are floating with varying degrees of intervention, and fiscal and monetary arrangements designed for a world of stable prices are crumbling under the near 20 per cent inflation being suffered by Britain and the lower but two digit inflations of other major countries.

Yet, in the face of these massive and hardly precedented developments, official (Treasury) and independent (e.g. the National Institute of Economic and Social Research) forecasters are attempting to predict and control the future course of the British economy using a model of economic behaviour which virtually ignores money and monetary phenomena; which makes forecasts of the real economy – unemployment and the rate of growth – independently of monetary developments, and which makes *ad hoc* guesses about the likely developments of wages and prices. The results of these forecasting methods have never been perfect but will, in 1975, be severely misleading.

Indications for 1975

The current forecasts on offer by the NIESR are for unemployment to rise but to stay well below one million during 1975, for real growth to be virtually nil and for inflation to increase further to 25

[103]

per cent. My own guesses are that by mid-year unemployment will hit the million mark with real income growth strongly negative. Inflation will remain in the high 'teens but will not (unless temporarily) reach the 25 per cent being forecast by the NIESR. The current account of the balance of payments will probably improve somewhat but not dramatically so.

The really serious developments in 1975 however will not be in the depth of the recession but in a continuing strong redistribution of wealth primarily from the old to the young. Such redistributions have already taken place on a massive scale in 1974. For example, through the building societies alone depositors have transferred, in real terms, some £1,700 million to house buyers. Losses on National Savings last year were of the order of £1,000 million.[1] A repeat of this in 1975 will reduce many people to levels of real wealth far below their expectations and lead to a great deal of strife and unrest.

Further, in 1975, despite the severity of the recession, there will be widespread shortages and bottlenecks in the supply of goods and services. This will further exasperate the average citizen and raise the degree of frustration, disappointment and even despair.

These are the broad features of the British economy which I foresee for 1975. What is the basis of these predictions? Why is the British economy in such a mess and what, if anything, can be done to improve matters?

International effects of oil prices

It is obvious that many of Britain's ills are shared by the other major countries, pointing to at least one source of our problems being international. The most obvious common source was the massive oil price increase inflicted by the OPEC in the autumn of 1973. A shock of this type is extremely difficult to absorb, not only because it lowers living standards but also because it creates an unusual degree of uncertainty and indecision and imposes heavy transitional adjustment costs particularly on producers but also on consumers. The classical case against monopoly is that its too high price and too low output misallocates resources because the value placed upon the marginal unit of the monopolised product is greater than the marginal cost

[1] With inflation at around 20 per cent and nominal net-of-tax yields on building society deposits and National Savings of about 10 per cent, losses on these assets are themselves of the order of 10 per cent. The figures quoted are based on that percentage loss.

of producing it. However, we must add to this the costs which are transitory but probably substantial which arise because, following a large *change* in the price of a monopolised product relative to other prices, many subsequent price *changes* and resource allocation *changes* are needed before the economy settles down at its new (monopoly distorted) equilibrium. Finding the new equilibrium is not simply a matter of calculating the oil content of every commodity and then increasing its price by the amount which maintains some 'desired' or 'normal' margin. Such a change may be implemented initially but the resulting changes in relative prices will lead to changes in consumption patterns which will in turn have implications for inventories and profits that will lead to further adjustments.

When a change as massive as that introduced by the oil price rise has to be absorbed it is inevitable that a great many mistakes will be made *en route* to the new equilibrium which will involve unusually heavy excess supplies (unemployment) and demands (bottlenecks) in various sectors. The market is a splendid instrument for responding to small changes introduced by the slow and gradual changes in tastes and technology but even the market cannot respond faultlessly to a shock of the proportions of the oil price rise. It would be a mistake to infer that because the market cannot respond instantly and correctly to such a shock that some form of government intervention is required. It is virtually certain that an attempt to find the new pattern of allocation and prices by non-market means would be even less efficient than the market itself.

In addition to the problems just outlined there is the problem of uncertainty about the durability of the OPEC cartel. Is it a permanent feature of the world or will it break? If it breaks, will we return to the old relative price of oil?

To make this discussion more vivid, consider the sorts of adjustments the economy has to make while searching for a new equilibrium pattern of resource use in just one oil-using activity, that of transport. With the new relative price of oil, what will be the effect on the demand for private and public transport? What will be the effect on the demand for cars of different carrying capacities and engine sizes? Qualitatively, it is easy to make the right predictions but quantitatively it is impossible. Car manufacturers and public transport agencies however have to make quantitative responses and have to invest in production systems based on guesses. There will be mistakes which result in over-production of some lines and under-production of others. Also inappropriate prices may prevail for a substantial period.

Throughout the adjustment process there will be a greater degree of re-allocation of labour and other productive resources than normal, leading to an increase in frictional unemployment. Eventually the economy will discover, by trial and error, how people's demand patterns have been affected by the new price structure. However, a lot of costly mistakes will have been made *en route*.

The last time the British economy had to respond to a shock as large as the oil price change was in 1925, when the 1914 gold parity was restored. This imposed on the economy the need for massive price and wage changes which caused a great deal of damage during the lengthy adjustment process. The miners' wages were at the centre of both these adjustments. In 1926 the equilibrium money wages of miners (and everyone else) clearly needed to fall. In 1974, the price of coal and the money wages of miners clearly needed to rise relative to other prices and wages. The cost of achieving these adjustments is clearly enormous.

Timing of reflation/deflation
A second factor affecting Britain at the present time which seems to be common to many countries is the timing in recent years of monetary and fiscal reflation and deflation. After tight monetary and fiscal policy in 1970 and 1971 all the major countries ran expansionary policies during 1972 and the first part of 1973. In the latter part of 1973 and throughout 1974, most major countries have been pursuing savage deflationary demand management policies. The synchronous expansionary policies of 1972-73 put the world on its present two-digit inflation rate and Britain on its present 20 per cent. The synchronous deflation of 1973-74 is leading to a world recession of proportions much worse than any experienced since the inter-war years. Without the oil price problem, the world (and Britain) would have a very deep recession in 1975. With the oil price adjustment problem remaining superimposed upon the world recession, the prospects are extremely gloomy.

Political staging of expansion and contraction
How has the world in general and Britain in particular got into this

[106]

synchronised expansion of contractionary monetary and fiscal policy? Two separate questions arise: first, why has there been a strong alternation between expansion and contraction, and second, why are those movements so strongly synchronised at the present time?

The alternations seem best explained in political terms. Governments believe, probably correctly, that low unemployment and rapidly rising real incomes are electorally advantageous. They also recognise that on the average over the life of a government they are powerless to affect these factors. They can, however, affect their distribution over time. By a tight reign on the economy immediately after an election, the government can create some slack in resources and have a moderating effect on inflation. In the run-in to the next election it can then stimulate demand, reducing unemployment, producing temporarily fast growth in real incomes and, because of the time-lags, avoid an excessive take-off in inflation. The inflation will hit the economy the year after the election and provide an excuse for the tight deflationary policies needed to create the slack for the next pre-election boom. In the 1970s these activities have been more closely synchronised across the major countries than they were in the 1960s.

In addition to these cyclical aspects of monetary and fiscal policy, there has been a tendency to aim for an average rate of unemployment and average growth rate which are more favourable than those which economic reality will permit. This has involved running the economy (not only the British) with periods of excessive demand pressure more prolonged and severe than the corresponding periods of demand restraint.

Further, there has been a persistent tendency to try to keep interest rates low. These factors taken together have resulted in money supply creation at rates which have tended upwards throughout the 'sixties and 'seventies to date.

The reasons for my predictions for the severity of Britain's 1975 recession will be coming clear. With all the major countries with whom Britain trades going into deeper recession in 1975 and with deflationary domestic policies being pursued, Britain is undoubtedly going to have a deepening recession this winter and spring at the very least. I suspect a deeper recession and sharper rise in unemployment than the NIESR and other forecasters largely because I

[107]

attach more importance to the powerful contraction of real money balances particularly in the USA, Germany and Britain and to the interactive effects of depression which I suspect are underestimated by conventional forecasting techniques.

My predictions about inflation in 1975 are based on the same considerations as those outlined above combined with the judgement, not shared by some, that there is still a tendency for prices to inflate less quickly in the face of generalised excess supply. With a very deep world recession basic commodity and raw material prices will fall and there will be less pressure making for price rises in all markets including those for labour.

Time-lags limit corrective action

I do not believe that much can be done to affect the course of real output, employment and prices in 1975 largely because of the time-lags in the effects of policy changes. We could do more to affect output than prices but even there, in view of the strong deflationary forces at work, the scope for corrective action is extremely limited.

What could be done, and in my view most emphatically should be done, is the introduction of limited, simple and specific measures designed to moderate or, preferably, eliminate entirely the most severe of the redistributive consequences of next year's inflation. This would be achieved by the issue of cost-of-living index-linked government bonds and national savings assets, and the freeing of the housing finance market to pay and charge realistic interest rates. It would also involve modifications to personal and corporate tax laws to ensure that *real* incomes and profits and not *nominal* incomes and profits were taxed. The specific changes needed in this respect are simple and well known.[1] The building societies should be encouraged to introduce an indexed mortgage option, the central feature of which would be to permit the spreading of the *real* burden of repayment of a mortgage equally over its life. At current inflation rates with equal *nominal* repayments, the *real* repayment dwindles to virtually nothing after a decade. This is not the place to rehearse the arguments for and against indexation. On this I can do no better than to refer again to Professor Friedman's IEA *Paper*.

[1] Milton Friedman, *Monetary Correction*, Occasional Paper 41, IEA, 1974.

Crucial ingredient – re-assignment of control instruments

The introduction of limited indexation can do much to offset the harmful effects of inflation but nothing to affect inflation itself. That requires longer-term policies. The crucial ingredient in such a policy is, in my view, the re-assignment of instruments of economic control. The post-war pattern has been to assign monetary demand management to the achievement of full employment; the use of the exchange rate (and a temporary import surcharge) to achieve external balance; the use of wage-price controls (with more or less severity) to achieve price stability. Demand management has been used in a cyclical fashion as indicated above and wage and price controls used from time to time as apparently required by the current rate of inflation. The change required is the assignment of monetary demand management to the eventual achievement of price stability and the development of labour market policies to minimise the unemployment rate.

The real danger is that, as most major countries run into recession next year and inflation begins to moderate, there will be an over-reaction in the reflation of demand which, in 1976 (in time for the US Presidential Election) will put the world into some degree of excess demand and, by 1977, to inflation rates which average in the low 20's with Britain somewhere in the high 20's or even 30's.

Extrapolating from past political behaviour, such a scenario is only too likely. To avoid it demand management is going to have to be much steadier, and the growth rate of the money supply gradually reduced until inflation and inflationary expectations have been squeezed out of the system.

My own reading of the political priorities is that we are not going to take the steps required; that the indexation issue is going to be evaded; that 1975 will be the biggest 'Stop' year we have had since the war; 1976 a bigger 'Go' year than hitherto and 1977 break new inflation records. The political mood of the country by then will be at best desperate and at worst mutinous.

[109]

A 'Manifesto' for 1975

SAMUEL BRITTAN

Principal economic commentator, 'Financial Times'
Visiting Fellow, Nuffield College, Oxford

IS THERE an imminent world economic catastrophe on the scale of the 1930s? I am rather sceptical. There is indeed a serious medium-term political and economic threat to freedom and democracy. This arises from (a) the generation of excessive expectations in the political market-place and (b) the rivalry of coercive monopolistic groups of which the unions are the main but not the sole examples. These dangers are intensified by (c) the chimaera of a 'just' or 'fair' reward for each of us in terms of status and money and (d) the alternative (and incompatible) widespread espousal of economic equality as a goal. Economists have here been especially at fault, in treating 'equality' almost as a commodity to be traded off against other commodities in compromise amounts, without always bothering to distinguish between the desire to help (or 'level up') the poor, the unfortunate and the unlucky, and simple envy of the rich, the successful, the lucky and the merely conspicuous.[1]

Confusion with wrong-headed prophecy

These medium-term warnings have, as was probably unavoidable, become hopelessly confused with largely wrong-headed prophecies about an imminent world economic catastrophe on the scale of the 1930s. Ever since I have been in economic journalism, hardly a year has passed without solemn predictions of a depression like that of the 1930s. The reasonable response: 'No; we face problems but not of that type', cannot hope to compete in news value. Moreover, like all negative assertions, it risks falsification by events. Nevertheless I shall stick to it.

[1] This argument was developed at some length in a Paper at the British Association Meeting in September 1974, which achieved some notoriety as a 'doomster's' utterance. I do not mind the notoriety if it causes some people to examine the positive ideas it contains. A revised version, 'The Economic Contradictions of Democracy', is to appear in the April 1975 issue of the *British Journal of Political Science*.

[110]

In the first place, whatever troubles we face in a world of double-digit inflation, they are of a different nature from those of the 1930s, when falling general price levels were a major world problem. Secondly, it is – to use a current cant political phrase – simply 'obscene' to compare even the worst of post-war recessions (the OECD is talking about a cumulative 4 per cent drop in American real GNP) with the 1930s slump when US output fell by over 40 per cent, when one out of every four American workers was *genuinely* out of work without today's cushioning arrangements, and when unemployment in Germany reached 6 to 7 million. Thirdly, the underlying world danger is still one of inflation; and the real danger of a new Great Depression comes from those who talk about it most, and who would stoke up the inflationary fires further by rushing into fresh excesses of deficit finance and money creation at the first signs of withdrawal symptoms from the inflationary disease.

The energy crisis is another fashionable myth. I well remember calling on the late Georg Tugendhat at the end of the 1950s during a similar scare about an 'energy gap' after Suez only to be told, 'The gap is in your mind'. Keynes referred more diplomatically to a common psychological trait, 'a certain hoarding instinct, a readiness to be alarmed and excited by the idea of exhaustion of resources'.[1] (A Middle Eastern War is a different matter. Although it could have grave economic repercussions, it would hardly be an economic crisis in origin; and, fortunately for me, even the most generous interpretation of my brief excludes consideration of Arab-Israeli diplomacy.)

As for the other aspect of the problem, the payment of the oil revenues to the producers, this is reminiscent of the 'transfer problem' associated with German reparations after the First World War. German opinion was almost unanimous in blaming their country's hyper-inflation of the early 1920s on reparations. Yet after the mark was stabilised in the winter of 1923, the 'transfer problem' magically dissolved and overseas investment in Germany over the following five years comfortably exceeded all reparations payments.

Not right 'cyclical' year for catastrophe: high inflation in late 1970s?
This brings us to the present UK situation. If we chart our position by reference to past cycles, this is the wrong year for inflationary

[1] Essay on Jevons in *Essays in Biography*, new edition, Macmillan, 1971, p. 117.

catastrophe. Unemployment and industrial slack are increasing. Despite the prospect of some individually alarming settlements, labour market conditions are not right for an acceleration of money earnings throughout the year; and the developing slack could still bring a balance-of-payments turn-round. Well before the end of the year alarm over unemployment may be expected to outrun inflationary fears; and, on past precedent, we should see a major injection of spending power while prices are still rising in double digits, thus posing a threat of German- or at least Brazilian-style inflation and political reaction, not now but for the later 1970s.

This is still the most probable prospect. But there are some major differences compared with the past. We are *as of this moment* entering a recession with a major *non-oil* deficit, which we did not have at this stage of any previous cycle. Moreover, there is the whole syndrome associated with the Stock Exchange collapse, business fears of 'Bennery', taxation of fictitious sums created by inflation, permanent price control, corporate failures and the effect on wage claims of the bipartisan desire to save every 'lame duck' in sight, and to guarantee all visibly threatened jobs. There is no precedent against which to assess the effect of these forces, how far they are real and how far wish-fulfilments on one side and nightmares on the other, and the true scale on which they might operate.

A fashionable view in some quarters is that the Government will be forced to adopt a wage freeze, which together with the Prime Minister's support for the EEC, will deprive it of the backing of 'the Left', and that a 'Government of National Unity' (GNU) will be formed. Anything is possible. But this vision seems to rest largely not on analysis, but on hope by Conservatives whose own irresponsible financial policies when in office were so largely responsible for our present troubles.

Futile expedients

The Government has a number of expedients left before it has to resort to wage controls. There is the widely advocated 'independent arbiter' to assess whether wage agreements violate the 'Social Contract'; there is compulsory notification of settlements, or the tightening up of the wording of the Social Contract itself, and bringing the CBI in on the act; there could even be a voluntary indexed wage-freeze once the miners are either out of the way or can clearly be

[112]

treated as a special case. The futility of these expedients need not prevent them from being tried, nor even buying time with international financial opinion during a time when cyclical forces may be working in the Government's favour.

Moreover, even if it does come to wage controls, or a retreat from the present extreme conceptions of full employment, what gain would there be to the Government from taking in Conservatives, Liberals or even independents? The new entrants would be blamed for all the unpopular measures and trade union hostility would be a hundred-fold more bitter and intense.[1]

Alternative policies

Nevertheless, it is not a waste of time to think of alternatives to current policies. For, despite what I have said about probabilities, a year in politics is a long time; and it would be unforgivable if anti-collectivists were to be caught without ideas if an opportunity for them arrived. Moreover, it is precisely those who are not in the political driving seats who could, if they knew where to go, set the terms and the issues of the national debate.

Unfortunately the liberal centre is bogged down in two issues. One is the supposed need for a wage freeze – which, whatever the case for it as a carefully timed adjunct to other policies, is almost certain to be botched up in practice – and a quasi-mystical involvement in the EEC, an unargued belief in which is regarded as the test of a person's political or even personal soundness. (I write as someone who believes, with the Foreign Secretary, that it would be folly to quit the EEC, who would vote 'Yes' in a referendum, and who is not blind to the sham toughness of renegotiations, which are being artificially protracted to provide some good theatre for the governing Party.)

How much unemployment?

Moreover the liberal centre and the 'radical right' are locked in a futile debate on the question 'How many unemployed would we

---

[1] The Government would, of course, be in parliamentary danger if the Left actually voted with the Opposition, or abstained when the Opposition was dividing the House on issues of confidence. But, although accidents can happen, I doubt if the parliamentary Left is so irrational and suicidal as to risk replacing the most promising Government from its point of view in the history of this country by a Coalition, a Conservative-dominated Government or a dissolution which would almost certainly see the annihilation of the Tribune Group.

need to conquer our present inflation?' The answer is largely a matter of leadership, personality, presentation, psychological atmosphere and general confidence. It is odd that politicians who normally stress precisely these aspects should be so insistent on a number which, in the nature of the case, cannot be provided even approximately.

A 'manifesto'

After all this, I ought to sum up with a few proposals of my own (which I would like to think of as a 'manifesto' for the extreme, or radical, as distinct from the insipid centre). For space reasons they will have to be concise and dogmatic; but I make no apology for starting with proposals to reform the political market-place, which is now the main weakness in our national life. On the one hand, a dominant minority of the electorate has too much influence. On the other hand, voters as a whole have too little choice. In no other country have electoral alternatives been confined to just two, and the same two, national leaders and to the same two (different but overlapping) personal styles ever since the early 1960s.

A. Constitutional and political framework for stability

1. To prevent fundamental changes being imposed by a government elected by a minority of the electorate, we should move to proportional representation (PR). The best method would be a German-type system retaining single-member constituencies, but with a margin of national seats to allot to parties which return less than their share of members.

2. Fixed four- or five-year Parliaments, with provision for emergency dissolution in rare, exceptional cases.

3. Party leaders should not serve more than two normal Parliaments, but need only submit themselves for re-election at the beginning of each Parliament (although a Members' meeting should be able to displace a leader at any time if a clear alternative commands a 'constructive majority').

4. As a first step to a written Constitution, a Bill of Rights guaranteeing *inter alia* editorial freedom, rights of union members and non-members, repealing the Official Secrets Act, guaranteeing the 'right to privacy', the right to choice outside and inside the

[114]

state system in health and education, abolition of corporal punishment in schools, to be replaced by sanctions such as 'exclusion', reduction of *compulsory* school-leaving age, etc.

5. If we are to have political and therefore economic stability there must be a time limit for withdrawal of British troops from Northern Ireland. Province to become independent and in charge of its own security. Boundary adjustment and population interchange with the Republic.

6. Stay in EEC, stop quibbling over Common Market budget and over (dubiously desirable) regional policy. On the other hand, state categorically that Brussels Commission is not embryo Government, Monetary Union to be shelved, oppose unnecessary harmonisation and campaign for reformed agricultural policy based on purchase in cheapest (*world*) market.

B. *Financial policy (macro-economic)*

7. As temporary second best to currency stabilisation, UK monetary and fiscal policy to give top priority to putting absolute ceiling on inflation at present rate. Then inflation-proof personal and corporate taxes, and issue indexed bonds. Encourage indexed mortgages and 'stable value' contracts.

8. Abolish price control as *quid pro quo* for not controlling wages.

9. Pending more fundamental measures on union monopoly power, stop subsidising strikes by transforming all social security payments to strikers' families into loans recoverable through PAYE. Tax rebates to strikers to be given no more quickly than to other citizens.

10. No tax changes, estimates or plans to be announced without corresponding expenditure figures on same basis; and no expenditure plans without corresponding projection of tax revenues.

11. Genuine float for pound; to be supported only by funds invested in London on commercial or interest-rate considerations.

C. *Far-reaching measures for efficiency (micro-economic)*

12. Set up an Adjustment Assistance Board in place of the National Enterprise Board (NEB). To provide *temporary* finance for

firms in trouble or their workers, assist break-up of insolvent enterprise into viable units, but *not* to perpetuate loss-making enterprises. To operate with fixed budget and eventually take in all Industry Act assistance, aerospace support and UK agricultural subsidies.

13. Revive Tax Credit Scheme as main instrument for redistribution, but taper-off credits for higher incomes to concentrate sum available at bottom end.

14. To improve housing market, make council houses transferable on short, long or perpetual lease at discretion of sitting tenant.

None of these measures is impossibly hair shirt. Some might even be popular, and most should be politically saleable given a modicum of leadership. They will not solve our problems permanently (nothing will). But they should see democracy through until after 1984.

I have *not* mentioned ration books, identity cards, utility furniture, the death penalty or a ban on pleasure motoring. Those whose ideal is a wartime nation with 'fair shares' for all in air-raid shelters and the barracks will doubtless go to other writers for their inspiration – there are dozens of nationalist Prussian Hegelians, most of them contemptuous of free markets and of consumer choice, waiting to be rediscovered.

1975 or 1984?

HARRY G. JOHNSON

Charles F. Grey Distinguished Service Professor of Economics,
University of Chicago

BRITISH ECONOMISTS should be thinking about the implications of their concentration on annual disasters, and the resulting encouragement of 'stop-go' policies, for the country's likely economic insignificance in 1984, rather than about the crisis of 1975. A crisis is highly probable, given the stop-go policies and the gathering retreat into increasingly ideological wishful-thinking of British policy discussion on the one hand, and the facts of life in the world economy on the other.

Ingredients of crisis

The ingredients of a crisis are to be found, proximately, in a new round of expansionism designed to reassure a badly frightened business community, a so-called 'social contract' that ensures inflationary concessions to the desire of the trade unions not to sacrifice expectations of continuing real income improvement themselves made possible in the past only by drawing on foreign resources through a balance-of-payments deficit of unsustainable dimensions, and the prospect of further increases in at least the nominal price of imported oil. Some part in the crisis may well be played by a change in Arab willingness to invest unusable monopoly profits in the London money markets.

A scenario for the crisis is difficult to advance with confidence; if one had the confidence, one would be too busy pyramiding one's financial bets to write it out, and too jealous of losing some potential profit to publicise one's analysis. However, the focus of the crisis is likely to be a sharp downward trend of the exchange value of the pound, resulting domestic inflationary wage and price movements, and a loss of foreign confidence leading to speculation against the pound (more accurately, growing recognition that the British authorities are as usual pretending that they have control over a

[117]

situation they have themselves rendered explosive). The touching-off is likely to be associated with recognition that the American authorities – by elephantine ponderosity rather than clever policy management – have broken the back of their inflationary boom.

Change policies or await crisis

There is no real question of leaving present government policies broadly unchanged, first because it is difficult to say what these policies are, beyond political opportunism, but second and more important because policies change only in response to crisis. Hence the short answer is that leaving present policies unchanged must lead to crisis.

Tactics and strategy are difficult to distinguish with reference to British economic policy; a more useful and relevant distinction is between tactics and catch-as-catch-can in the absence of strategy. Given the way policy thinking has been evolving in recent years, the obvious change to expect is a statutory income policy or indefinite wage freeze plus the introduction of direct balance-of-payments controls over imports. In other words, the likeliest changes are escalation of the class war, disguised as restraining the greed of the workers and punishing their ingratitude for the benefits of full employment conferred by their educated betters, and further self-inflicted wounds for the faltering British economy.

Let Hong Kong run Britain?

I do not see how 'the danger of worsening crisis' can be overcome, since no one really wants to avoid it. Crisis is too useful to politicians and the so-called economists who toady to them in the hope of selling their own panaceas, to be given up so lightly. If the British government would surrender its sovereignty for, say, the next ten years, and let the government of Hong Kong administer the United Kingdom without fear or favour (and, to avoid contamination, without any governmental visiting in either direction), the situation might possibly be straightened out.

[118]

IEA Publications

Subscription Service

An annual subscription to the IEA ensures that all regular publications are sent without further charge immediately on publication – representing a substantial saving.

The cost (including postage) is £10.00 for twelve months (£9.50 if by Banker's Order) – £7.50 for teachers and students; US $25 or equivalent for overseas subscriptions.

To: The Treasurer,
    Institute of Economic Affairs,
    2 Lord North Street,
    Westminster,
    London SW1P 3LB

Please register a subscription of £10.00 (£7.50 for teachers and bona fide students) for the twelve months beginning.................................

☐ Remittance enclosed   ☐ Please send invoice

☐ I should prefer to pay by Banker's Order which reduces the subscription to £9.50.

Name ....

Address ....

       ....

Position ....

Signed ....

Date   ....

OP43

[119]

IEA OCCASIONAL PAPERS in print

EATON PAPERS in print